BIOCHEMISTRY RESEARCH TRENDS

OLEIC ACID: DIETARY SOURCES, FUNCTIONS AND HEALTH BENEFITS

BIOCHEMISTRY RESEARCH TRENDS

Additional books in this series can be found on Nova's website under the Series tab.

Additional e-books in this series can be found on Nova's website under the e-book tab.

BIOCHEMISTRY RESEARCH TRENDS

OLEIC ACID: DIETARY SOURCES, FUNCTIONS AND HEALTH BENEFITS

LUCIANO PAULINO SILVA
EDITOR

NOVA BIOMEDICAL
New York

Library of Congress Cataloging-in-Publication Data

Library of Congress Control Number: 2013934506

ISBN: 978-1-62618-332-2

Published by Nova Science Publishers, Inc. † New York

CONTENTS

PREFACE

Oleic acid is a monounsaturated fatty acid ubiquitous in the nature. Due to its intriguing physical-chemical and biological properties, there is growing interest in the oleic acid as an agricultural product, a nutritional and health-promoting component, an industrial feedstock, and even a renewable energy resource. The main objective of this book is to present readers with comprehensive information on oleic acid. This book has 11 chapters. Chapter 1 describes the importance of oleic acid under an evolutionary perspective. Chapter 2 describes the natural variation and potential enhancement of oleic acid in oilseed crops. Chapter 3 is devoted to compare volume and adsorption properties of oleic acid solutions with other C18 fatty acids. Chapter 4 describes microbial transformations of oleic acid. Chapter 5 describes the use of heterogeneous catalysis for the conversion of oleic acid into biodiesel. Chapter 6 provides readers with vegetable sources, technological and health implications of oleic acid. Chapter 7 describes the oleic acid content in milk of different mammalian species. Chapter 8 provides information on the importance of oleic acid in brain tissue. Chapter 9 describes the status and therapeutic importance of oleic acid present in vegetable oils on human health. Chapter 10 describes some applications of oleic acid on enzyme immobilization. Finally, Chapter 11 describes roles of oleic acid in cancer. I am indebted to participating authors who provided authoritative views on different aspects of oleic acid and I invite readers to join us in this endeavour.

In: Oleic Acid: Dietary Sources, Functions and Health Benefits ISBN: 978-1-62618-332-2
Editor: Luciano Paulino Silva

Chapter 1

EVOLUTIONARY ASPECTS OF OLEIC ACID BIOSYNTHESIS AND INCORPORATION FROM UNICELLULAR TO MULTICELLULAR ORGANISMS

Cínthia C. Bonatto[1] and Luciano P. Silva[2,*]
[1]Embrapa Genetic Resources and Biotechnology,
Brasilia, Brazil
[2]University of Brasilia, Institute of Biological Sciences,
Brasilia, Brazil

ABSTRACT

The idea of life's origin in the context of lipid membranes and fatty acids can be considered uncontroversial. Considering the unique physical-chemical properties of oleic acid, it is not difficult to understand its biological importance. Despite its cytotoxicity against several cell types, there are powerful health benefits and biotechnological applications of oleic acid. Therefore, it is not surprisingly that oleic acid is ubiquitous in nature. A systematic literature compilation has been undertaken on the oleic acid relative content of more than one thousand of living organism species belonging to all the major taxonomic groups. In comparison with other C18 fatty acids, oleic acid is the most abundant type among all biological Domains, but among biological Kingdoms, Plantae is the unique in which oleic acid is not the most abundant C18 fatty acid. The simultaneous use of oleic acid and other C18 fatty acids' content was sufficient to distinguish and hierarchically cluster main taxonomic groups in accordance to common and reliable phylogenetic approaches. In light of these facts, the use of oleic acid and other C18 fatty acids' content could effectively benefit taxonomic inferences and even represents a novel outlook of biological diversity and evolution.

[*] E-mail address: luciano.paulino@embrapa.br, lucianopaulinosilva@gmail.com

INTRODUCTION

In the Biology, abiogenesis is the field that deals the origin of the life from nonliving to living matter resulting from a long process of molecular evolution started around 3-5 billion years ago [1]. Despite this fact, from an anthropocentric viewpoint it is interesting to sight that some major building blocks of the life such as amino acids [2], fatty acids [3], and nucleic acids [4] were first discovered only between one and a half and two centuries ago. Billions of years before human scientific knowledge, the formation of simple organic molecules, like fatty acids and amino acids may have allowed the essential building blocks required to engineering complex molecules like structural lipids and proteins, respectively.

In the lipids scenario, it is possible that oleic acid and some of its cognate C18 homologous fatty acids such as stearic, linoleic, and linolenic acids have implications to the origin of the life on prebiotic Earth as major structural components of potential protocells such as micelles [5], reverse micelles [6], surfactant-coated oil droplets [7], and mainly vesicles formed by monolayer [8] and bilayer [9] membrane structures. This fact per se justifies the growing interest in research on C18 fatty acids. Overall, as exhibited in Figure 1, scientific literature reveals a considerable growing in C18 fatty acids science last few years. However, whereas the interest in oleic acid has grown exponentially over the last decade, the interest in other C18 fatty acids has grown almost linearly for this period. Therefore, it is now accepted that oleic acid is a true protagonist in the fascinating functional role of fatty acids on biological processes. Since oleic acid elicits important biological responses, the aim of the present chapter was to review from an evolutionary viewpoint the literature related to oleic acid in comparison to its cognate C18 fatty acids in unicellular and multicellular organisms.

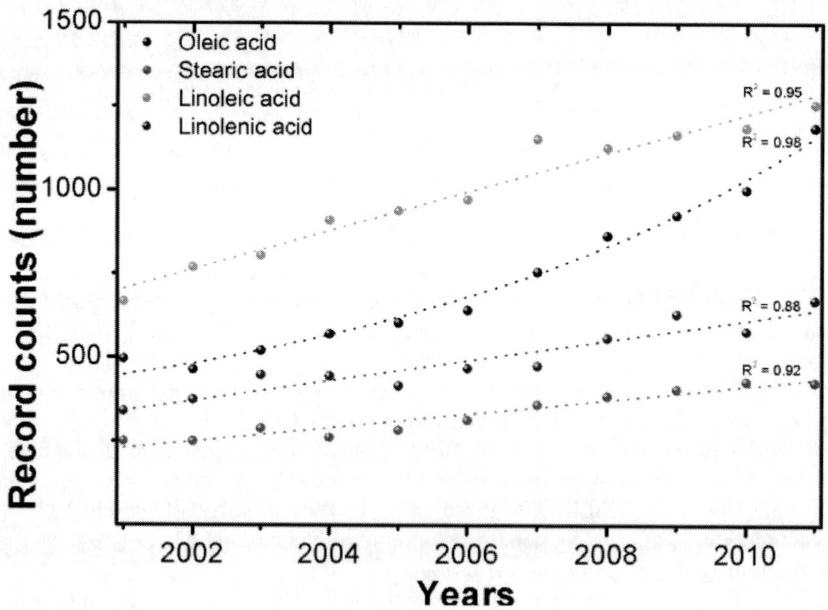

Figure 1. Annual number of C18 fatty acids related reports assessed as scientific research papers published since the 2000s as obtained from ISI Web of Science database. Dashed lines represent linear fits for stearic, linoleic and linolenic, and exponential fit for oleic acid data.

OLEIC ACID AND THE FATTY ACIDS SCENARIO

It is largely accepted that fatty acids are essential components for all known Earth life forms. Unsurprisingly, live organisms share at least some functions attributed to fatty acids like mechanical structure, source of energy, cellular signaling, and insulation. Fatty acids are chemically defined as carboxylic acids with a long aliphatic chain showing variable length and number of double bonds. Depending on the absence or presence of double bonds between each two carbon atoms in fatty acids chain, they may be structurally termed saturated or unsaturated, respectively. In spite there are many systematic nomenclature for fatty acids, several academics use the nomenclature recommended by the International Union of Pure and Applied Chemistry (IUPAC) [10]. The IUPAC system names fatty acids exclusively based on the number of carbon atoms linked together, and the number (if any) and position of each double bonds (Z or E, also referred as cis or trans) counted from the carboxyl group. Contrarily to the precise, but nontrivial IUPAC nomenclature, fatty acid names have been subjected to shorthand notations. All these notations adopt the form C:DB, where C is the number of carbon atoms and DB is the number of double bonds along the aliphatic chain. Unsaturated fatty acids are also commonly reclassified into families or series according to the position of the first double bond relative to the methyl end of the molecule (omega, ω or n). In addition, the delta (Δ) system is similar, but not only the first, but all double bond positions are specified as well as their specific configuration (cis or trans). Oleic acid is probably the most common and abundant fatty acid in nature. The biosynthesis of oleic acid occurs by desaturation of stearic acid (previously coupled with CoA) mediated by the action of the iron-dependent enzyme stearoyl-CoA 9-desaturase with NADPH as a cofactor [11]. Chemically, stearic acid is delta9-cis desaturated to oleic acid that in some organisms may also be converted by additional desaturation or even elongation steps into polyunsaturated fatty acids like linoleic acid, linolenic acids and longer chain fatty acids [12]. Oleic acid is a monounsaturated (monoenoic) omega-9 cis(Z)-fatty acid with the chemical formula CH3(CH2)7CH=CH(CH2)7COOH (C18H34O2), has the IUPAC name (Z)-octadec-9-enoic acid (CAS Number: 9000-69-5) and has a simplified molecular input line entry syntax (SMILE) notation CCCCCCCC\C=C/CCCCCCCC(O)=O. The chemical structure of oleic acid is showed in Figure 2. It has a monoisotopic molecular mass of 282.25 g/mol; density of 0.895 g/cm3; melting point of 16.3°C; boiling point of 286°C at 100 mm Hg and 360°C at 760 mm Hg; viscosity of 25.6 cP (30°C) and 4.85 (90°C) mPa·s; molar volume of 313.897 cm3; topological polar surface area of 37.30 Å2; 15 rotatable bonds; molar refractivity of 87.063 cm3; refractive index of 1.4585 at 26°C; enthalpy of vaporization of 66.53 kJ/mol; pKa of 5.02 in aqueous solution at 25°C; surface tension of 32.8, 27.94, and 21.6 dyne/cm at 20, 90, and 180°C, respectively; vapor pressure of 5.46x10-7 mm Hg at 25°C; specific heat of 2.046 (50°C) J/g; iodine number of 89.9; acid value of 189.6; combustible when exposed to heat or flame; flash point of 189°C (closed cup); autoignation temperature of 363°C according to ChemSpider [11] and Hazardous Substances Data Bank (HSDB) [12]. Although it is insoluble in water, but soluble with ethanol, ether, acetone, benzene, chloroform, carbon tetrachloride, it may also be soluble in aqueous media as a salt (e.g. sodium), ester, or anion to give oleate micelles or vesicles [12]. After more than one century of its discovery, oleic acid synthesis was first performed in 1934 when Noller and Bannerot published the seminal paper claiming a synthetic compound that was believed to be identical with the natural [13].

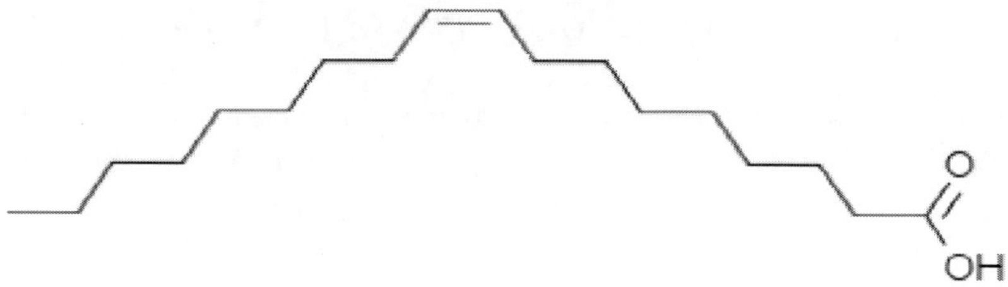

Figure 2. Typical representation of the chemical structure of a *cis*-oleic acid molecule.

Oleic acid has at least ten double bond positional (*cis*-6 to 16) and additionally 15 configurational (*trans*-2 to 16) natural 18:1 isomers which are mainly produced during partial hydrogenation of more unsaturated fatty acids [14]. In addition, oleic acid can also be converted to other fatty acids by natural or synthetic processes [15].

Some of these molecules are represented as ideal models in Figure 3. Since free fatty acids (e.g. oleic acid) are hydrophobic due to the single hydrocarbon long chain, they are water-insoluble. Anyway, major biological processes in which fatty acids display crucial roles are hydrophilic in physical-chemical nature.

Due to this reason, it is not surprisingly that most naturally occurring fatty acids (e.g. oleic acid) are esterified to form more complex lipids that contain one, two or more additional chemical moieties and include mono-, di-, and triacylglycerols; cholesterol, ergosterol, and phytosterol; phospholipids; glycolipids; and proteolipids.

Figure 4 shows typical oleic acid-derived lipids also represented as typical ideal models, including typical phospholipids commonly used as major building blocks of lipid bilayers. Lipid bilayers are semipermeable supramolecular systems that allow the compartmentalization of cellular chemical reactions.

According to their packing states, lipid bilayers can be grouped into at least three lamellar phase behaviors: crystal, gel, and fluid that can be seen as molecular dynamics models in figure 5. Such states are related not only to structural features, but also to the functional aspects of the bilayers.

Despite the significance of ideal models as prototypes of the fatty acid 3D structures and their derivatives, X-ray crystallographic investigations have observed a great repertoire in the molecular conformation of fatty acid chains, in particular oleic acid [16, 17].

Such amazing feature can be easily observed when comparing the plethora of conformational variants of oleic acid in complex with different proteins (Figure 6). Note that although it is recognized that fatty acids rapidly and passively diffuse through lipid bilayers; the uptake, translocation, transport, and utilization of fatty acids in unicellular and multicellular organisms may also be mediated and eased by proteins [18]. Albumins [19], lipid transfer proteins [20], fatty acid binding proteins [21], and lipases [22] are only a few examples of proteins exhibiting fatty acid association properties. Although this molecular plasticity has been demonstrated, several significant questions remain almost all related to how these conformational changes influence fatty acids function and even potential adverse effects.

Oleic acid is cytotoxic at micromolar *in vitro* concentrations towards bacteria [23], fungi [24], and mammalian cells including B-lymphocytes [25], T-lymphocytes [26], epithelial cells [27], and tumor cells [28, 29] in spite several positive health effects have been well-documented.

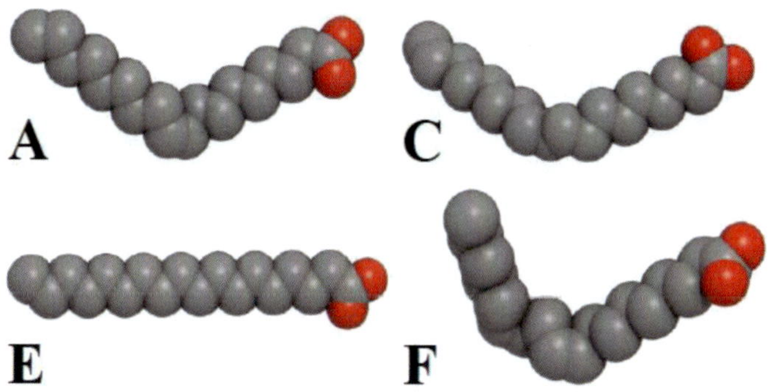

Figure 3. Ideal 3D molecular representations of oleic acid and some of its C18 homologues displayed as realistic QuteMol models using PDB atomic coördinates available on Hetero-compound Information Centre (Uppsala). (A) (*Z*)-octadec-9-enoic acid (oleic acid), (B) (*E*)-octadec-9-enoic acid (elaidic acid), (C) (9*Z*,12*R*)-12-hydroxyoctadec-9-enoic acid (ricinoleic acid), (D) octadecanoic acid (stearic acid), (E) (9*Z*,12*Z*)-octadeca-9,12-dienoic acid (linoleic acid), and (F) (9*Z*,12*Z*,15*Z*)-octadeca-9,12,15-trienoic acid (alpha-linolenic). The models were colored by atom type (carbon-gray and oxygen-red) and the hydrogen atoms were not displayed for clarity.

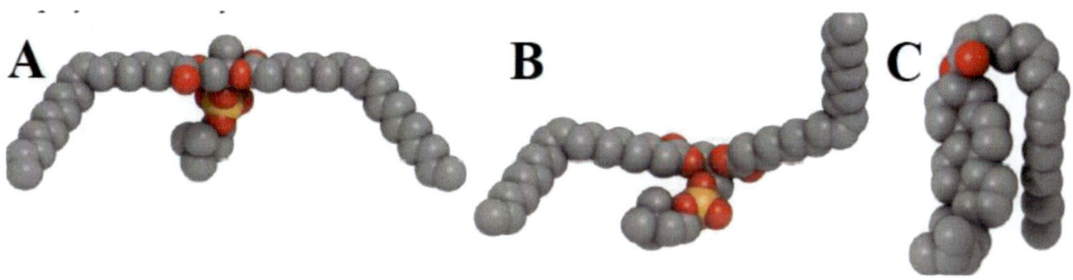

Figure 4. Ideal 3D molecular representations of complex lipids derived from oleic acid displayed as realistic QuteMol models using PDB atomic coördinates available on Hetero-compound Information Centre (Uppsala). (A) dioleoyl-phosphatidylcholine, (B) 1,2-dioleoyl-*sn*-glycero-3-phosphocholine, (C) cholesteryl oleate. The models were colored by atom type (carbon-gray, oxygen-red, and phosphorus-orange) and the hydrogen atoms were not displayed for clarity.

Figure 5. Three-dimensional molecular dynamic models of 1-palmitoyl-2-oleoyl-sn-glycero-3-phosphocholine (POPC) bilayers in different phases displayed as realistic QuteMol models using PDB atomic coördinates available on Dr. Helmut Heller webpage (http://www.lrz.de/~heller/membrane/membrane.html). (A) crystal, (B) gel, (C) fluid fully hydrated phases. The models were colored by atom type (carbon-gray, oxygen-red, and phosphorus-orange) and the phospholipids hydrogen atoms were not displayed for clarity.

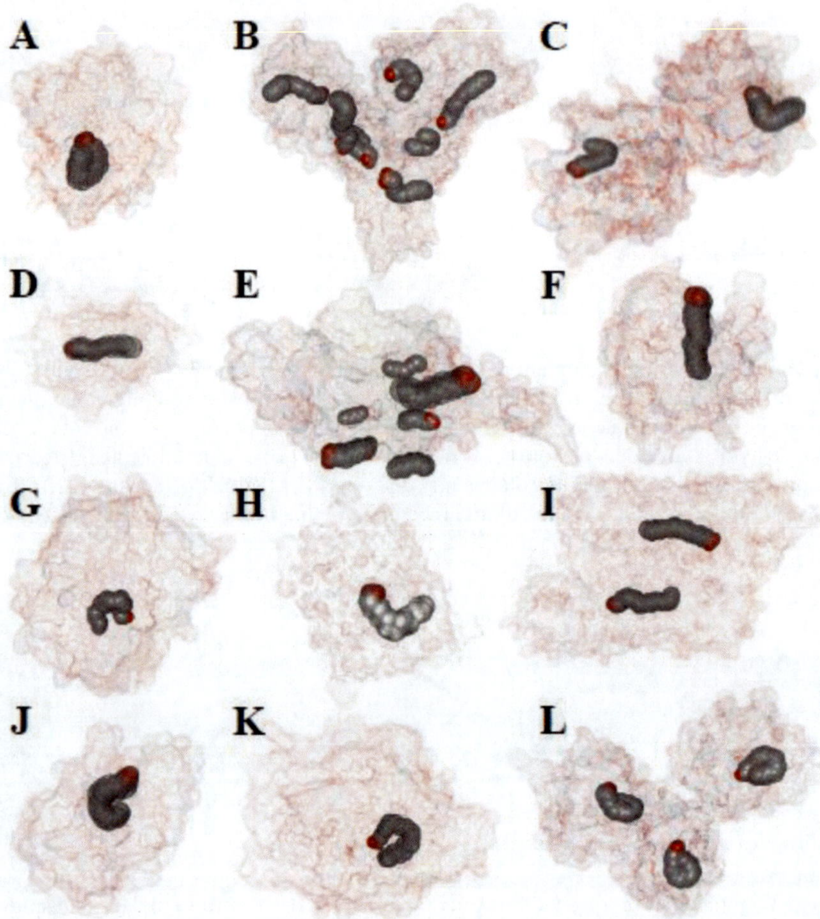

Figure 6. Three-dimensional X-ray crystallographic structures of proteins in complex with oleic acid molecules displayed using BALLView software as SES and VDW surfaces, respectively, from PDB atomic coördinates available on Protein Data bank (PDB). (A) *Schistosoma mansoni* fatty acid binding protein (1VYF), (B) human serum albumin (1GNI), (C) S146A mutant of *Thermomyces* (Humicola) *lanuginosa* lipase (1GT6), (D) maize lipid transfer protein (1FK5), (E) *Chlamydomonas reinhardtii* channel rhodopsin light-gated cation channel (3UG9), (F) bovine beta-lactoglobulin (4DQ3), (G) mice adipocyte lipid binding protein mutant (1G74), (H) *Locusta migratoria* fatty acid binding protein from flight muscle (2FLJ), (I) regulatory domain of the *Mycobacterium tuberculosis* adenylyl cylcase, at ph 6.0 (2EV1), (J) *Ambystoma mexicanum* liver bile acid-binding protein (2FTB), (K) human muscle fatty acid binding protein (1HMS), and (L) bovine cellular lipophilic transport protein (1PMP). The models were colored by atom type (carbon-gray, oxygen-red, and phosphorus-orange) and the hydrogen atoms were not displayed for clarity.

Health benefits of oleic acid *in vivo* include lowering the risks of heart and circulatory system diseases [30], diabetes [31], and cancer [32]; helping in weight loss [33]; accelerating wound healing [34]; and reducing the inflammation [35].

Industrial chemistry applications of oleic acid include production of biofuels and biolubricants [36]; excipient in pharmaceuticals and cosmetics [37]; emulsifying agent in aerosols [38]; preserving foodstuffs [39]; fixing textile pigments [40], delivering immunomodifying compounds [41]; formulating detergents and soaps [42]; and protective coatings for metals [43].

OLEIC ACID AND THE ORIGIN OF THE LIFE

It is widely accepted the notion that in a prebiotic scenario of the primitive Earth simple molecules underwent events of self-assembly towards formation of complex biological molecules which culminated with the origin of the life around four billion of years ago [44]. Simple fatty acids may have been protagonist components of early cell membranes by encapsulating other essential biological compounds like nucleic acids (RNA and DNA) and amino acids and finally forming protocells [45].

However, it is noteworthy that only some decades ago the concept of protocells was nicknamed [46] and the recent field of the biological sciences termed synthetic biology was not capable to re-sculpt the life yet [47]. Despite this fact, it has been demonstrated the growth and division of protocells of small unilamellar oleate vesicles supplied with fatty acid micelles and leading the encapsulation of RNA molecules mimicking a primitive genome [48]. It is a prebiotically plausible example since the substance used to build such vesicle is a salt of oleic acid that can be synthesized in aqueous environment.

In addition, heterogeneity of shape, size, and even fatty acid composition could be related to the origin of the life and such aspect could lead to the first competitive environments. In fact, protocell vesicles can undergo competitive growth when in the presence of another type of vesicles in a contest similar to the negative interaction that occurs among organisms requiring the same limited resource [45]. Therefore, the study of protocell models based on oleic acid and other fatty acids may contribute to a better understanding of the life from the ancestral survivor protocell lineage to the beginning of the biological evolution and biodiversity.

OLEIC ACID AND BIODIVERSITY

The classification of living organisms into clusters is an ancient human approach to spread biological information. It is plausible imagine that human populations have always discriminated between edible or poisonous organisms to communicate desirable or undesirable features, respectively. The biological field that deals with identification and classification of the organisms in terms of similarities or dissimilarities is termed taxonomy.

The use of morphological and physiological features is well established in classical taxonomy to assign organisms into appropriate biological groups aiming the construction of a tree of life (Figure 7). However, there is growing debate on the role of molecular methods can act in species identification and even how chemical constituents of a species or part of it can become the major means to cluster unequivocally species.

In fact, molecular barcodes involving the use of standard markers of DNA [49], proteins [50], and fatty acids [51] has demonstrated to be enough to distinguish among groups of organisms and in some situations species or even components of them, like organs in multicellular organisms. Although fatty acids profile has been used for decades to distinguish species [51], literature references regarding approaches compiling such information under the form of databases are scarce and restrict to comparisons among bacteria [52], yeast [52], molecular structures [53], plant seed oils [54], or foods [55]. In addition, most of these databases are available only for private uses and require expensive commercial licenses.

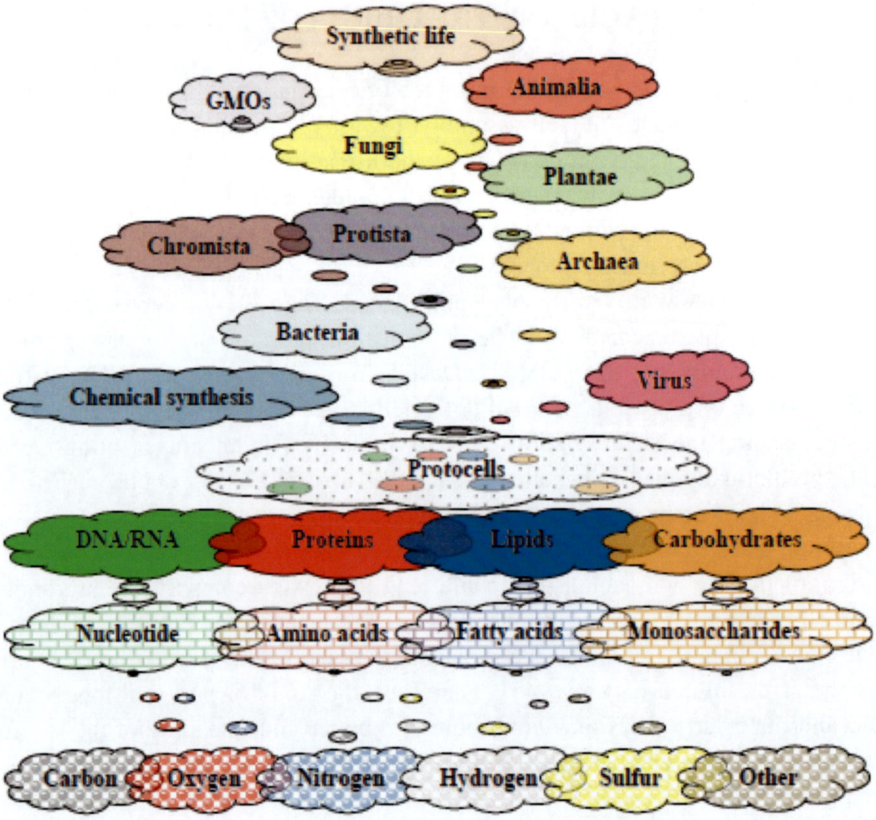

Figure 7. Schematic representation of the tree of life from essential building blocks to manmade interventions. GMOs, genetically modified organisms.

The present chapter handles with a literature search aiming to perform a high-throughput comparison of C18 fatty acid profiles and mainly oleic acid (18:1) relative content observed among unicellular to multicellular organisms (n=1034) belonging all major groups of the life, some membrane-bounded viruses (n=10), mitochondria (n=7), chloroplasts (n=8), and some mammalian diseases related to oleic acid (n=12). The fatty acids composition of unicellular organisms and small invertebrates were obtained from total lipids of whole cells or organisms. Otherwise, for most multicellular organisms, several authors reported differences among the fatty acid profiles of organs that could reflect specific requirements of plants and animals [56, 57]. Leaves are the major site of chemical energy production by plants. Otherwise, skeletal muscles are the major site of chemical energy consumption by vertebrates. Therefore, in the present chapter, plant leaves and vertebrate muscles were the organs selected for comparisons.

Researchers have used a variety of methods described in the literature to determine quantitatively the fatty acids composition of the organisms or parts of them. These analytical methods include liquid [59], gas-liquid [59, 60], and thin-layer [61] chromatographies. Despite the significant increase in resolution provided by some contemporary techniques, it is still not possible to resolve some specific minor fatty acids, including isomers. Due to this fact, double bond positional and configurational isomers were grouped together in the present chapter. Therefore, the C18 fatty acids were identified only as 18:0, 18:1, 18:2, and 18:3.

The use of fatty acid contents for taxonomic resolution is problematical and controversial since medium [62], diet [63], pathologies [64] or environmental differences [65] could directly reflect on such profiles. However, some authors claim the use of such approach under native conditions of the organisms to perform the chemotaxonomy of a group of organisms. The major aim of the present chapter is check the potential use of C18 fatty acids to distinguish high-level taxonomical groups' instead specific species.

All data regarding C18 fatty acid profiles of organisms belonging to a particular group were obtained from available tables described in papers and book chapters previously published and these data were shown in the present chapter as mean ± standard error of the mean. All comparisons were performed at $P < 0.05$ (one-way ANOVA - Tukey's test) using the statistical analysis freeware PAST [58]. In addition, taxonomical groups were analyzed using multivariate statistics to identify similarities and/or dissimilarities among them (cluster analysis using Euclidean distance).

Oleic Acid and Biological Domains

The Three Domain System is a biological clustering that groups all living organisms into three different taxonomic levels: Bacteria, Archaea, and Eukarya [66]. Bacteria and Archaea members are prokaryotes (unicellular organisms that lack a cell nucleus and membranous organelles) and Eukarya members are eukaryotes (unicellular or multicellular organisms that have a cell nucleus and membranous organelles).

Despite this dichotomy, this hierarchical approach is mainly based on the differences among the 16S rRNA gene of the organisms. However, from our knowledge of the literature in this area, no studies have attempted on global comparison of fatty acids profile of organisms belonging these Domains.

Figure 8A shows a comparison of the C18 fatty acids composition among the three Biological Domains. Oleic acid was the most abundant fatty acid in all Domains, but all three groups differ statistically among them ($P < 0.05$) with Eukarya showing 75% higher levels of 18:1 than Bacteria and 61% lower than Archaea that has 4.5-fold higher content than Bacteria. All other C18 fatty acids were more abundant in Eukarya than in Bacteria or Archaea ($P < 0.05$) and 18:2 was not observed in Archaea. Although the purpose of the present chapter is not to scrutinize other fatty acids, it is interesting to observe that the sum of the relative content of all other fatty acids is also different among the Domains ($P < 0.05$).

To rationalize these data based on a phylogenetic perspective, C18 fatty acids (18:0, 18:1, 18:2, and 18:3) and the sum of all other fatty acids relative contents were hierarchically grouped using cluster analysis (Figure 8B). There is not a consensus regarding the common ancestor among these Domains, the evolutionary relationships remain unclear. Anyway, the present results corroborate the profound molecular differences observed among the Biological Domains that led to several authors adopt the Three Domain System as the highest taxonomic level of the life tree. In addition, these results suggest that Bacteria are closer relatives of Eukarya than of Archaea from the viewpoint of the C18 fatty acids content ($P < 0.05$). However, it is important take into account that those differences may be influenced by the sample size of each taxon, in particular the small number of species with available fatty acids profile for Archaea.

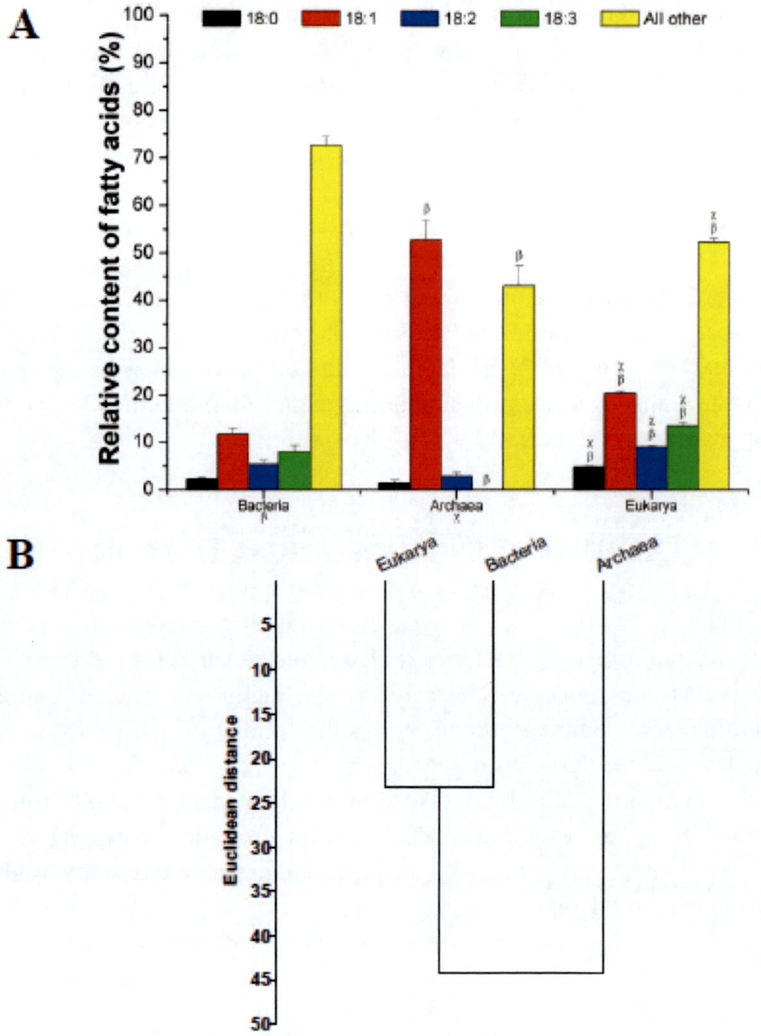

Figure 8. Relative content of C18 fatty acids in total lipids of typical prokaryotic (Bacteria and Archaea) and eukaryotic (Eukarya) organisms. All Life Domains are represented. (A) Histogram bars show mean ± standard error of the mean. α and β symbols represent statistically significant differences ($P < 0.05$, one-way ANOVA - Tukey's test) from Bacteria (n=107) and Archaea (n=80), respectively, when compared to Eukarya (n=847). (B) Cluster analysis (Euclidean distance) of the fatty acids data using unweighted pair-group average (UPGMA) algorithm.

Oleic Acid and Biological Kingdoms

Within the tree of life, Kingdoms are the second level of hierarchical clustering and are included in some of the three Domains. There is no consensus among researchers regarding what is the correct number of Kingdoms of the life. In the present chapter the former Kingdom Protista was divided into Chromista and remaining Protista for all comparisons. Kingdom (Domain) Bacteria includes only the previously termed subkingdom Eubacteria (true bacteria) whereas the Kingdom (Domain) Archaea includes the previous termed subkingdom Archaebacteria (ancient bacteria).

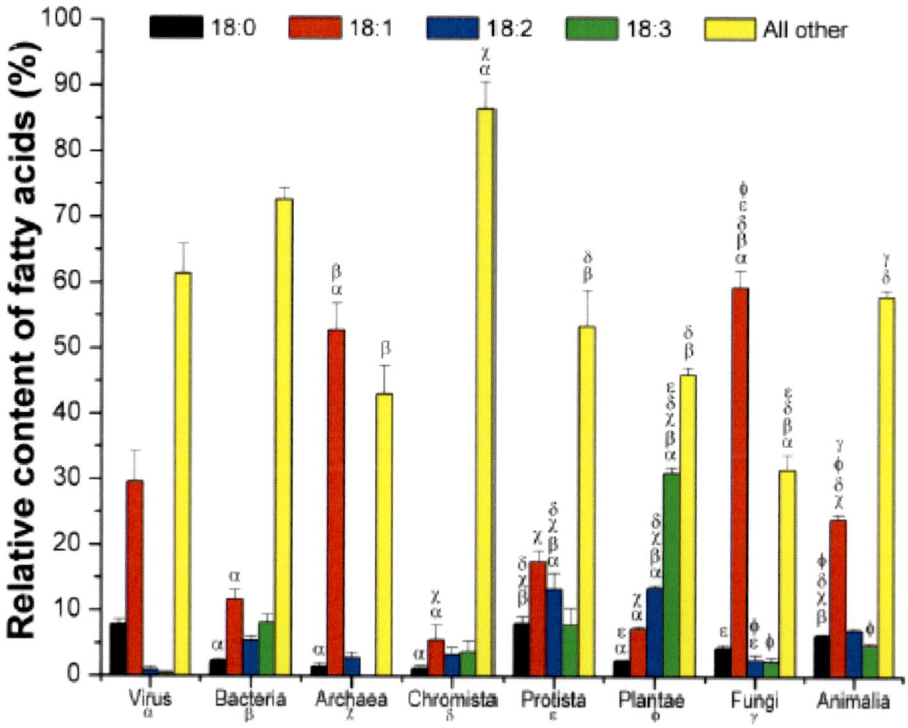

Figure 9. Relative content of C18 fatty acids in total lipids of typical virus, prokaryotic (Bacteria and Archaea), and eukaryotic (Chromista, Protista, Plantae, Fungi, and Animalia) organisms. All Life Kingdoms are represented also including virus, and splitting Chromista and Protista into two groups. Histogram bars show mean ± standard error of the mean. α, β, χ, δ, ε, ϕ, and γ symbols represent statistically significant differences ($P < 0.05$, one-way ANOVA - Tukey's test) from Virus (n=10), Bacteria (n=107), Archaea (n=80 isolates), Chromista (n=8), Protista (n=23), Plantae (n=286), Fungi (n=58), respectively, when compared to Animalia (n=472).

In addition, the membrane-bounded viruses were included as non-living chemical systems compartmentalized by lipid membranes. Figure 9A shows the summary of C18 fatty acids profile among all Biological Kingdoms and these quantitative results will be divided and discussed in detail into subsections for each taxon along the chapter.

Oleic Acid and Bacteria

Bacteria Domain is a non-monophyletic group, previously termed Kingdom Eubacteria. This group possesses currently about 10,358 described species [67]. Most Bacteria that have been described to date shows maximum diameters ranging from 0.5 to 5 μm; are prokaryotic organisms, as characterized by the lack of a membrane-enclosed nucleus; predominantly unicellular, but some species form colonies; show DNA in a single circular chromosome; and if it is present, the cell wall consists of peptidoglycan.

The Bacteria morphology can be spherical (cocci), rod-shaped, bacillary, or spiral. These microorganisms can be heterotrophic, autotrophic or chemotrophic and the nutrition pathways are predominantly by diffusion through cell wall and membrane. The process of reproduction are asexual (binary division or fragmentation) or sexual (conjugation). Bacteria Domain includes several microorganisms of medical and economical importance among which are *Staphylococcus*, *Lactobacillus*, *Salmonella*, *Escherichia*, and *Streptomyces*. They are also classified according to cell wall structure as Gram-positive or Gram-negative.

Overall, 18:1 is the most abundant C18 fatty acid in Bacteria (Figure 9) [68-82]. However, the proportion of 18:1 is significantly lower in Bacteria when compared to Virus, Archaea, and Fungi ($P < 0.05$). In addition, the proportions of 18:0, 18:2, and 18:3 are also lower in Bacteria when compared to Virus, Protista, and Animalia; Protista and Plantae; and Plantae, respectively ($P < 0.05$). Otherwise, taking into account the sum of all other fatty acids, Bacteria shows higher relative content than Archaea, Protista, Plantae, and Fungi ($P < 0.05$).

At present, members of Bacteria are formally clustered into about 23 phyla including Cyanobacteria [68-72], Proteobacteria [73-76], Firmicutes [77-80], Chlamydiae [81], and Spirochetes [82]. Figure 10 illustrates the overall data of cluster analysis of C18 fatty acids of Bacteria phyla. Based on this rooting, there is evidence that the profiles relationship within Bacteria is a continuum with different phyla diverging from a common ancestor in the order (Firmicutes (Cyanobacteria (Chlamydiae (Proteobacteria + Spirochaetes)))). These results are very similar to those obtained by other phylogenetic approaches including 16S+23S rRNA sequences [83] and Hsp60 and Hsp70 sequences [84].

Firmicutes shows the most distant taxonomic relationship among these Bacteria members. Firmicutes are gram-positive Bacteria in contrast to all other phyla compared in this chapter that are Gram-negative members. Moreover, Spirochaetes and Proteobacteria are sisters, something already observed by other authors using distinct methodologies [83, 84]. From these findings, insight is gained into the use of fatty acids to discriminate Bacteria phyla and even to bring out specific features like Gram-negative or Gram-positive types.

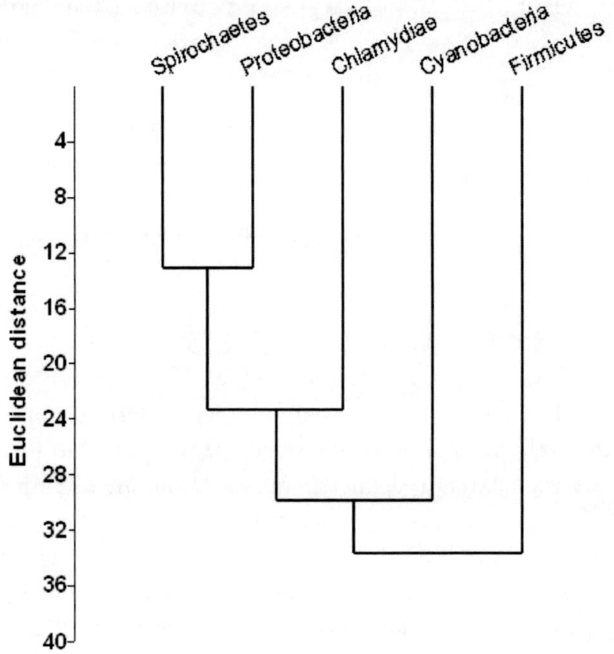

Figure 10. Cluster analysis (Euclidean distance) of C18 fatty acids relative content of common Bacteria phyla using unweighted pair-group average (UPGMA) algorithm available on PAST [56].

Oleic Acid and Archaea

The Domain Archaea is a group of unicellular prokaryotic microorganisms previously classified within Bacteria. The Archaea term was first introduced by Woese and Fox [85]. This group possesses currently about 502 described species [67]. The microorganisms are prokaryotes, unicellular, have a cell wall, plasma membranes, 70S ribosomes, flagella, and a single chromosome. The morphology can be spherical, bacillary, spiral, flat, square, discoid, or often irregular, and in several species there is pleomorphic morphology. Archaea dimensions are extremely variable with maximum diameters ranging from 0.1 to 15 μm, with some reaching filamentary forms with even 200 μm. The nutrition is predominantly based on absorption (diffusion), and the process of reproduction of the archaeas is asexual and occurs by binary division or fragmentation. Among fatty acids, 18:1 is the most abundant in Archaea [86-102] and the second highest among the Biological Kingdoms only comparable to Fungi contents (Figure 9). In fact, the proportion of 18:1 is significantly higher in Archaea when compared to Virus, Bacteria, Chromista, Protista, Plantae, and Animalia ($P < 0.05$). Only trace amounts (< 3%) of 18:0 and 18:2 are observed in Archaea. It is important to mention that archaeal cell membranes are mainly structured by molecules strongly different from those of all other known life forms including unusual isoprenoid fatty acids and that normal fatty acids constitute only a minor fraction that was renormalized in the present chapter.

Oleic Acid and Protista (Protozoa)

The members of the Kingdom Protista or Protozoa are commonly aerobic unicellular eukaryotic organisms, occasionally forming colonies or a single multicellular tissue sharing cells with similar or specialized functions, respectively. This group possesses currently about 8,118 described species [67]. They can be autotrophic (photosynthetic) or heterotrophic (heterotroph or organotroph). Heterotrophs take their food by phagocytosis and autotrophs perform photosynthesis and may have the chlorophylls a, b or c. They reproduce sexually (gametes) or asexually (binary fission).

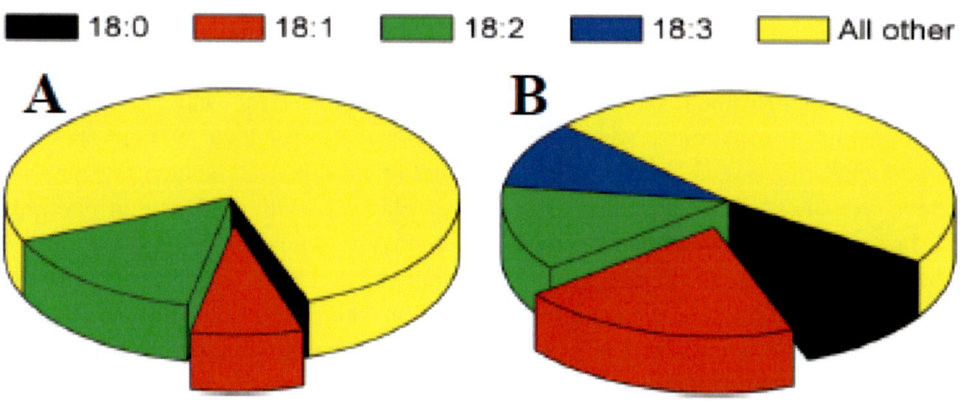

Figure 11. Relative content of C18 fatty acids in total lipids of ciliated (A, n=5) and flagellated (B, n=18) Protista organisms.

Similarly to Bacteria and Archaea, 18:1 is the most abundant C18 fatty acid among Protista (Figure 9), although palmitic acid (16:0) is the most abundant among all fatty acids (data not shown) [103-107]. Protista movement is achieved by several different modes including the use of cilia and flagella. Moreover, it is noteworthy to note that there are substantial differences on C18 and other fatty acids between ciliated and flagellated Protista probably due to locomotion needs (Figure 11).

OLEIC ACID AND CHROMISTA

The Kingdom Chromista is a eukaryotic group, probably polyphyletic, which has been treated as a separate Kingdom or included in Protista. This group possesses currently about 13,033 described species [67]. Microorganisms belonging to this Kingdom are unicellular, flagellated (two flagella), and have a cell wall. They can be chemoheterotrophs (parasites and saprobes) or photoautotrophs (photosynthetic organisms which have chloroplasts containing chlorophylls a and c which were acquired by endocytosis of intracellular secondary red algae). The chromistas can be found in two forms, a diploid (sporophyte) and haploid (gametophyte), which alternate throughout the life's cycle. During meiosis the diploid organism produce two different haploid zoöspores which undergo mitosis and produce gametes that can be isogamic, anisogamic and oogamic; since fertilization occurs there is a new diploid organism. In agreement with other published experimental data suggesting significant differences between Chromista and Protista allowing them to be clustered into distinct groups, the present chapter using compiled literature data show that the C18 fatty acids relative content in Chromista are 2- to 7-fold lower than those observed for Protista [108-109]. These findings reinforce the idea that these organisms must be grouped independently.

OLEIC ACID AND PLANTAE

The Kingdom Plantae is one main group of the life on Earth. This kingdom currently has about 215,644 described species [67], distributed among algae, bryophytes, pteridophytes, gymnosperms, and angiosperms. Plants are eukaryotic organisms, sessile, unicellular or multicellular, with or without specialized tissues, cell-walled, and they have life histories that show alternation of generations where there is a life's cycle that involves both sporophytic and gametophytic phases, although the latter is evolutionarily reduced in the most species.

Plants are categorized into nonvascular (lower plants, including algae and bryophytes) and vascular groups (higher plants, including pteridophytes, gymnosperms, and angiosperms), and the latter into seedless vascular plants (pteridophytes) and seeded plants (gymnosperms and angiosperms). They are photoautotrophs (perform photosynthesis) and this is possible due to the presence of chloroplasts which contain chlorophylls. Plantae members are key components of the production of organic matter in the marine, freshwater, and terrestrial habitats. Therefore, they are the first trophic level in the food chain that sustains the energy pyramid of the life. Beside foods (for example, grains, fruits, and vegetables), plant products vital to humans include wood and wood-derived products, fibers, drugs, oils, latex, pigments, and resins. Coal and petroleum are also fossil substances of plant origin. There is a common sense that Plantae fatty acids are usually unsaturated [110-117]. This fact may also be observed taking into account only C18 fatty acids since the relative content of 18:0 < 18:1 < 18:2 < 18:3 (Figure 9). The prevalence of highly unsaturated (more than two unsaturations) fatty acids over the lowly unsaturated and saturated ones in Plantae cells and tissues are due to unique biosynthetic pathways and enhances the fluidity of the cell membranes. Plantae members synthesize *de novo* 18:1 from 18:0 and several other unsaturated fatty acids, including very long chain polyunsaturated fatty acids, which may also be further synthesized by additional desaturation and/or elongation steps of 18:1 [118]. Among algae fatty acids content, there are only minor significant changes ($P < 0.05$) in Rhodophyta (red algae) and Phaeophyta (brown algae) when compared to Chlorophyta (green algae) (Figure 12). These findings are not surprisingly taken into account that such groups of lower plants share a common ancestry. Still among the lower plants, Bryophyta 18:1, and 18:1 and 18:2 fatty acids relative content than Phaeophyta, and Phaeophyta and Rhodophyta, respectively, and a lower 18:3 fatty acids relative content than Chlorophyta (Figure 12, $P < 0.05$). However, note that there are several significant differences ($P < 0.05$) in the relative contents of C18 fatty acids between lower and higher plants (Figure 12). Cluster analysis of the data grouped the plants into two major clades according to their vascular states (Figure 13). In addition, this chemotaxonomical approach reinforced the idea that Bryophyta originates from Chlorophyta [119]. In addition, it is interesting to observe that despite the gymnosperms and angiosperms share only a common ancestor around 300 million years ago, their C18 fatty acid profiles are similar ($P < 0.05$) and such characteristic could be related to a consequence of their common molecular features, including some major lipid metabolic pathways.

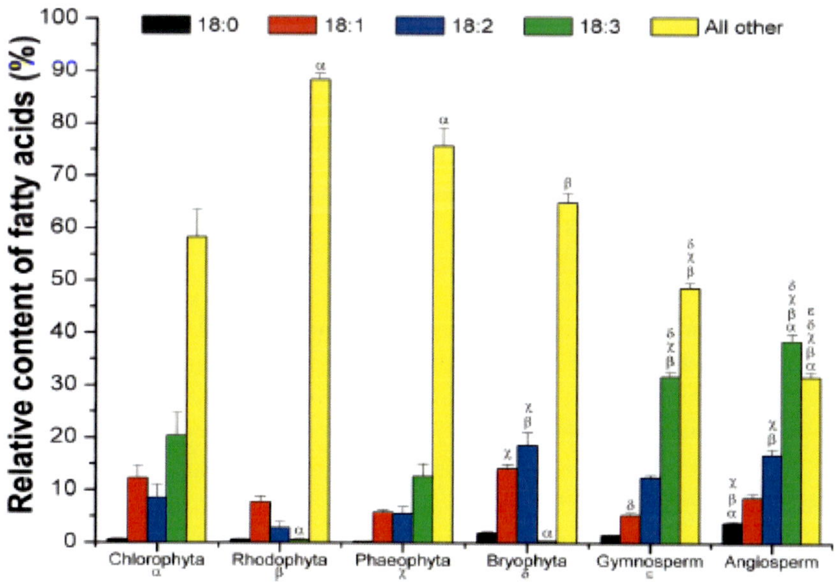

Figure 12. Relative content of C18 fatty acids in cell (Chlorophyta, n=8; Rhodophyta, n=17; and Phaeophyta, n=6) or leaf (Bryophyta, n=7; Gymnosperm, n=143; and Angiosperm, n=105) total lipids of Plantae organisms. α, β, χ, δ, and ε symbols represent statistically significant differences ($P < 0.05$, one-way ANOVA - Tukey's test) from Chlorophyta, Rhodophyta, Phaeophyta, Bryophyta, and Gymnosperm, respectively, when compared to Angiosperm.

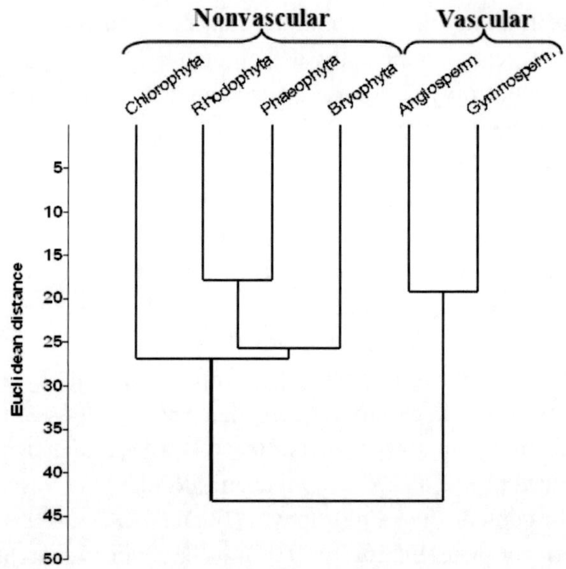

Figure 13. Cluster analysis (Euclidean distance) of C18 fatty acids relative content of common Plantae groups using unweighted pair-group average (UPGMA) algorithm available on PAST [56].

Oleic Acid and Fungi

The term Fungi was first used in referring to the Kingdom by Whittaker b in 1969. This group possesses currently about 43,271 described species [67]. These organisms are eukaryotes, heterotrophic, non-photosynthetic, have a cell wall (glucans, mannans and/or chitin), plasma membranes, DNA, ribosomes of the 80S type, and have vacuoles.

The morphology can be yeasts (rounded cells), mycelium (filamentous cells or not) or dimorphic fungi. The nutrition is predominantly by absorption, and the process of reproduction of the Fungi members is by either asexual (budding or binary fission) and sexual. The fungi members grow in a variety of habitats, including extreme environments such as desert, and high salt or high ionizing radiation areas, as well as in deep sea sediments. Others can survive even after intense UV and cosmic radiation encountered during space travels.

Fungi showed significantly higher 18:1 relative content ($P < 0.05$, Figure 9) than almost all other groups compared in the present chapter, except for sharing a similar content with Archaea [120-124]. The high 18:1 relative content of Fungi (in average more than 50%) All other differences on relative content of C18 fatty acids ($P < 0.05$) were observed only when Fungi were compared with Protista and Plantae and such attribute was also shared with almost all other life forms (Figure 9).

Fungi encompass a large variety of morphological attributes and life styles, in particular reproduction and locomotion statuses that were also reflected in the C18 fatty acids classification (Figure 14).

According to other reports using morphological or molecular [125, 126] approaches, C18 fatty acids also showed that Ascomycota and Basidiomycota are sister clades commonly classified as subkingdom Dikaria.

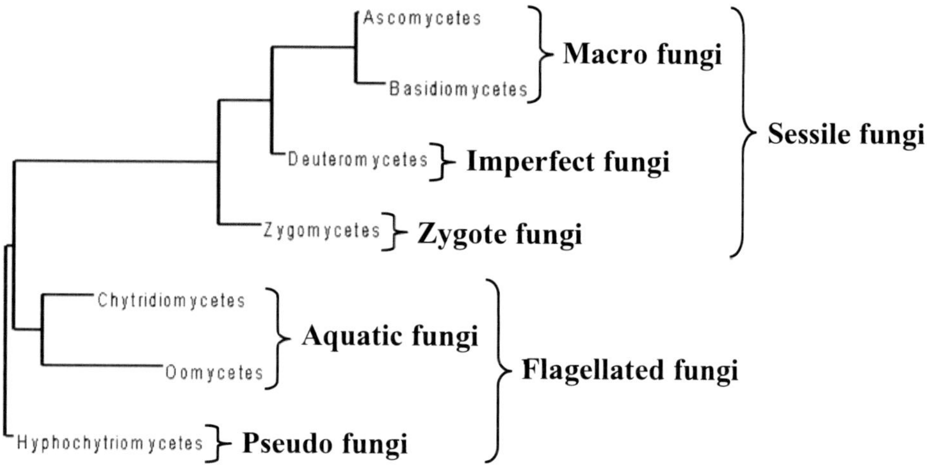

Figure 14. Neighbor joining cluster analysis (Euclidean distance) of C18 fatty acids relative content of common Fungi Classes using out-group rooting available on PAST [56].

Note that the artificial group Deuteromycetes which it was created to cluster all conidial fungi that do not have any connection with Dikaria, in fact formed a distinguishable clade using the present C18 fatty acids approach. The (Deuteromycetes (Ascomycota and Basidiomycota)) classes are in turn distinguishable from Zygomycetes in spite all they are filamentous fungi without motile stages. Additionally, this clade formed by sessile fungi is clearly separated from the clade formed by flagellated fungi belonging Chytridiomycetes, Oomycetes, and Hyphochytriomycetes classes. It is also worth noting that Hyphochytriomycetes was the out-group clade corroborating an array of evidence that sometimes classifies this group within Protista Kingdom.

Oleic Acid and Animalia

Over time the Kingdom Animalia has been divided into invertebrates and vertebrates reflecting a human need to make groups and to favor their own relatives throughout the history of humanity. Similarly to almost all Kingdoms (except Plantae), 18:1 is the most abundant C18 fatty acid (Figure 9).

Unsurprisingly, the Kingdom Animalia is the biggest one and currently has about 953,434 described species [67] distributed between invertebrates and vertebrates groups. From C18 fatty acids viewpoint there are only small significant differences between invertebrates [127-137] and vertebrates [138-168], in particular 18:0 and 18:3 relative contents (Figure 15). Since the relative levels of 18:1 fatty acid was similar between vertebrates and invertebrates ($P < 0.05$), it is possible that this fact shows the conservation of at least some metabolic aspects of the Kingdom Animalia, including biosynthetic and diet incorporation aspects. Because this fact, in the present chapter both major groups of Kingdom Animalia were divided into subgroups (phyla and classes) so that specific considerations could be properly performed. The invertebrate animals are eukaryotic, multicellular, and heterotrophic living organisms; have organized tissues (except Porifera); and reproduce sexually or asexually.

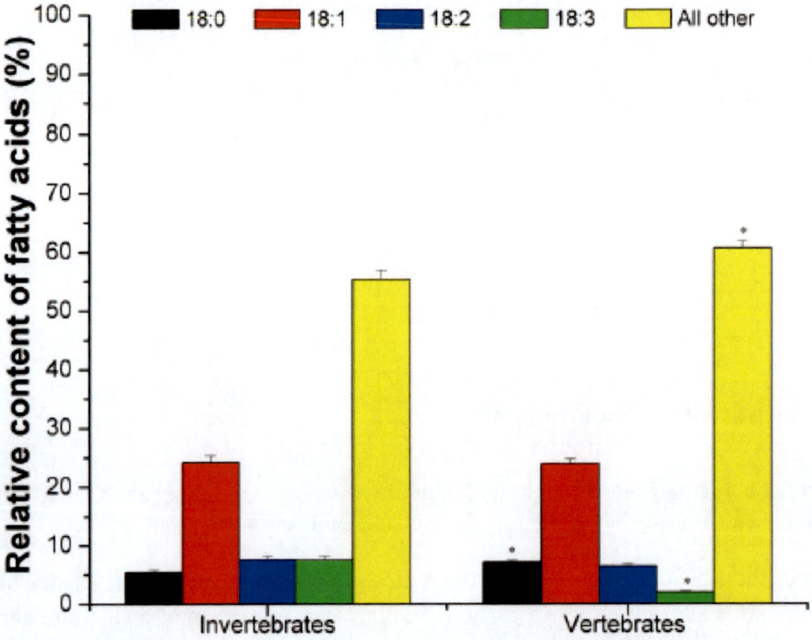

Figure 15. Relative content of C18 fatty acids in whole specimens (Invertebrates, n=238) and skeletal muscle (Vertebrates, n= 234) total lipids of Animalia organisms. Histogram bars show mean ± standard error of the mean. * symbol represents statistically significant differences ($P < 0.05$, one-way ANOVA - Tukey's test) from invertebrates.

Invertebrates are the most abundant macroscopic living organisms on all Earth's habitats. This paraphyletic group currently has about 895,432 described species, distributed among 20 known phyla, Porifera, Cnidaria, Ctenophora, Platyhelminthes, Nemertea, Gnathostomulida, Rotifera, Annelida, Pogonophora, Sipuncula, Echiura, Priapulida, Pentastomida, Onycophora, Tardigrada, Arthropoda, Monoblastozoa, Trilobozoa, Archaeocyatha, Proarticulata and three subphyla of the phylum Chordata: Uruchordata, Cephalochordata, and Hyperotreti, each one of which has structural peculiarities. From the fatty acids viewpoint, the invertebrate phyla are different in C18 fatty acids relative composition (Figure 16). As expected from the evolutionary relationships of the invertebrate groups, Porifera and Cnidaria share similar C18 fatty acids profile, mainly when compared to other invertebrate phyla ($P < 0.05$). However, the limited number of available studies describing the fatty acids profile of some invertebrate phyla (e.g. Echinodermata and Platyhelminthes) does not allow firm any unequivocal conclusions. The vertebrate animals are eukaryotic, heterotrophic, and multicellular living organisms; have organized tissues; and reproduce sexually. The term vertebrates is due to the presence of a serially arranged vertebrae along the axial endoskeleton (cartilaginous or bone). The vertebrates are composed of a set of basic systems, including integumentary, skeletal, muscular, digestive, circulatory, respiratory, excretory, reproductive, endocrine, and nervous systems. This group includes most of the animals belonging to the chordates phylum, and possesses currently about 58,002 described species. The vertebrates group is divided into five major subgroups; fish [138-151], amphibians [152], reptiles [153-155], birds [156-159] and mammals [160-168], each one of which has morphological, physiological, metabolic, and biochemical peculiarities, including specific C18 fatty acids content (Figure 17).

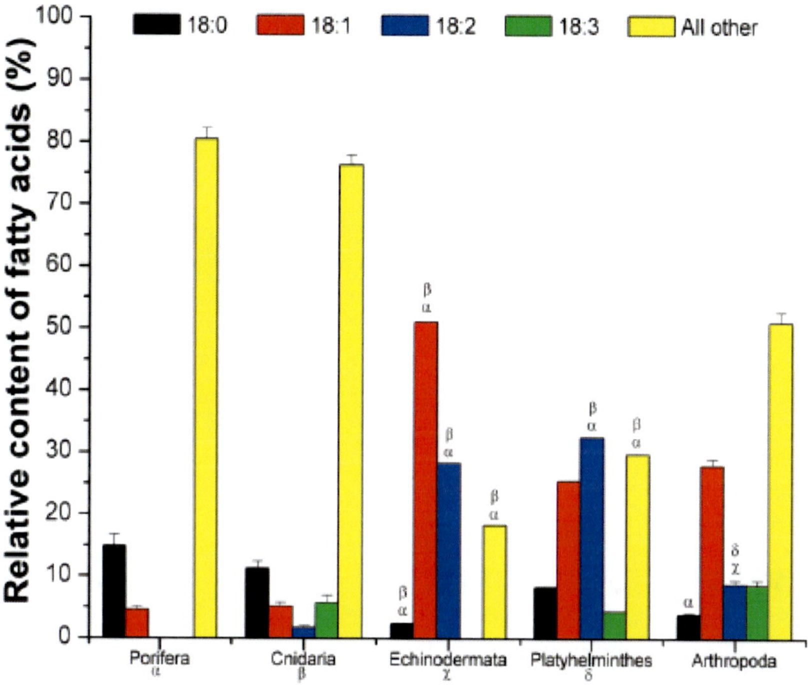

Figure 16. Relative content of C18 fatty acids in total lipids of typical invertebrate phyla. Histogram bars show mean ± standard error of the mean. α, β, χ, and δ symbols represent statistically significant differences ($P < 0.05$, one-way ANOVA - Tukey's test) from Porifera (n=24), Cnidaria (n=15), Echinodermata (n=1), and Platyhelminthes (n=1), respectively, when compared to Arthropoda (n=197).

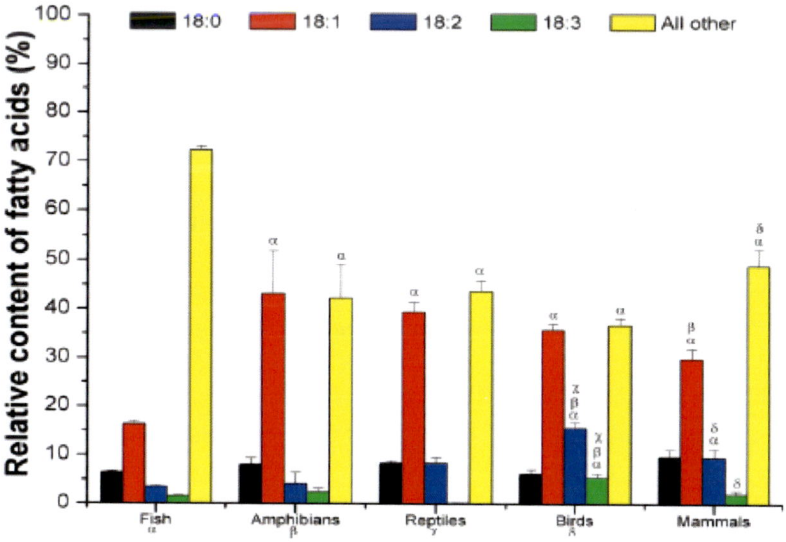

Figure 17. Relative content of C18 fatty acids in total lipids from skeletal muscles of the most common vertebrate groups. Histogram bars show mean ± standard error of the mean. α, β, χ, and δ symbols represent statistically significant differences ($P < 0.05$, one-way ANOVA - Tukey's test) from Fish (n=141), Amphibians (n=4), Reptiles (n=30), Birds (n=30), respectively, when compared to Mammals (n=29).

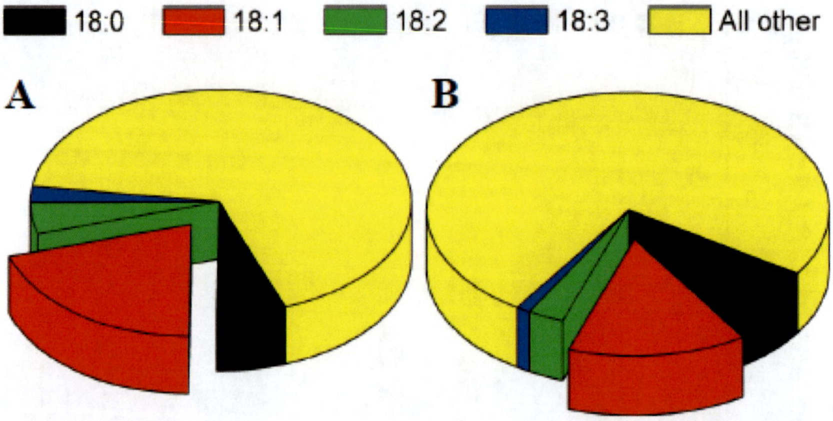

Figure 18. Relative content of C18 fatty acids in total lipids from skeletal muscles of freshwater (A, n=48) and saltwater (B, n=93) fishes.

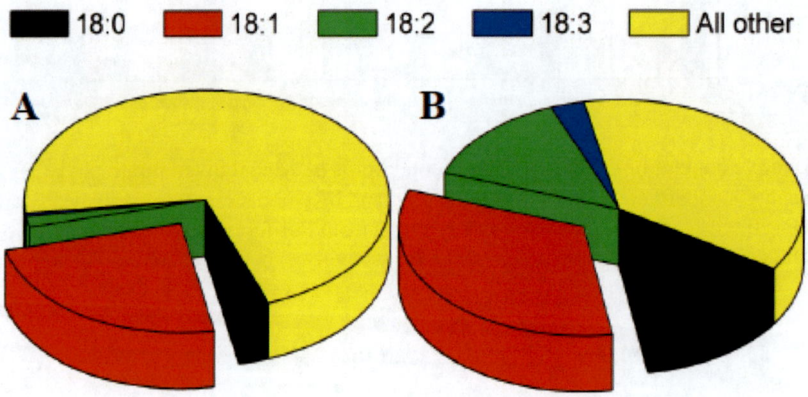

Figure 19. Relative content of C18 fatty acids in total lipids from skeletal muscles of aquatic (A, n=10) and terrestrial (B, n=19) mammals.

Fish is the unique vertebrates' subgroup for which all other fatty acids sum, in particular polyunsaturated fatty acids, is more abundant than the C18 fatty acids that contribute to a lower 18:1 relative content than any other vertebrates' class. Unsurprisingly from the evolutionary viewpoint, amphibians and reptiles share C18 fatty acids similarities (Figure 17). Birds have higher 18:2 and 18:3 contents than all other vertebrates ($P < 0.05$). Unlike fish (lower) and amphibians (higher), about one-third of mammal fatty acids content is 18:1 ($P < 0.05$). In addition, despite sharing the same bodies' temperature regulation system (endothermic), mammals and birds show consistent differences in some fatty acids content ($P < 0.05$).

Compared to saltwater (marine) fish, freshwater fish have higher 18:1 content (Figure 18) and this fact has been associated to the dietary fat [145]. Overall, saltwater fish species contain higher levels of polyunsaturated fatty acids. Since most aquatic mammals' lives in the oceans (marine) and some of their important dietary components include high polyunsaturated fatty acids saltwater fish, it is not surprisingly that they share similarities in fatty acids profile. In contrast, due to the lower amounts of dietary polyunsaturated fatty acids, this was not found in the terrestrial mammals (Figure 19).

OLEIC ACID AND MITOCHONDRIA

Mitochondria are membranous rod-shaped organelles responsible for cell respiration and finally energy transduction inside the cell. This organelle ranges from 0.5 to 10 μm in diameter. Morphologically the mitochondria are constituted by an outer membrane, an intermembrane space (the space between the inner and outer membranes), an inner mitochondrial membrane, the area of crests (formed by invagination of the inner membrane), and a matrix (area within the membrane internal). Mitochondria have their own DNA and essential components to self-replicate. In this sense, mitochondria are presumably organelles derived from endocytosis of primitive bacteria. The establishment of this exchange was favorable to the two bodies due to the cooperation of both: one of each offering protection and nutrients; and the other one in a compensatory way, favoring a higher income and energy used through the process of cellular respiration. From the evolutionary viewpoint based on nucleic acids sequence data, mitochondria are considered to have originated from Proteobacteria and the current use C18 fatty acids content corroborates this idea [169-170].

OLEIC ACID AND CHLOROPLAST

Chloroplasts are the photosynthetic organelles, present in Plants, some Bacteria, Archaea, and Chromista. This organelle ranges from 4 to 7 μm in diameter. Their ultrastructure consists of a thylakoid membrane that contains the molecular components required to convert light energy into chemical energy; and a stroma that contains the enzymes that use the NADPH and ATP synthesized by thylakoid to transform carbon dioxide into carbohydrates.

The chloroplasts have their own DNA and structure to self-replicate. There are several types of chloroplasts that differ in shape, ultrastructure and pigmentation. However, rRNA sequences and several other genes show a monophyletic source for all chloroplasts that have arisen from an endossimbiotic event between a eukaryotic cell (host cell) and a Cyanobacterium. The comparative use of chloroplasts C18 fatty acids in a cluster analysis approach [170-173] reinforces this hypothesis (Figure 20).

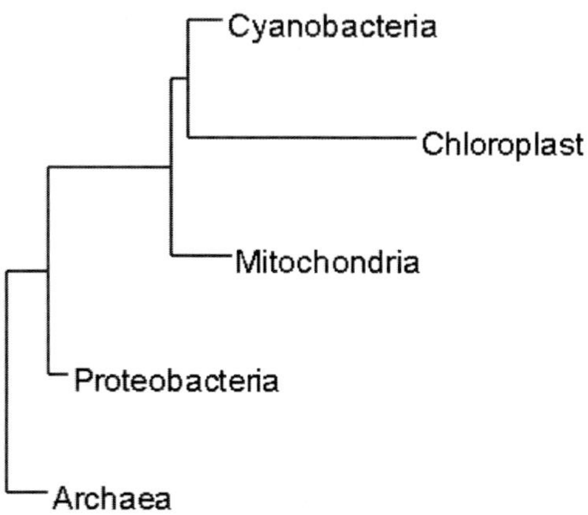

Figure 20. Neighbor joining cluster analysis (Euclidean distance) of C18 fatty acids relative content of mitochondria and chloroplast using out-group rooting available on PAST [56].

Oleic Acid and Virus

All known viruses are obligate intracellular parasites since they require an intracellular environment of a living organism to reproduce and specific viruses can infect virtually any organism types such as Plants, Animals, Bacteria, and Archaea. Most viruses that have been described to date share sizes ranging from 20 to 300 nm and are composed of nucleic acid molecules (DNA and/or RNA) which can be single or double strand; a protein envelope; and in some rarer cases by an additional envelope consisting of lipids similar to those observed in cellular membranes and obtained during viruses replication inside the cell. Due to this fact, it is widely accepted that fatty acids content, including C18 fatty acids (in particular 18:1), of viruses are indistinguishable from those observed in their host cells [174-175].

Conclusion

C18 fatty acids are ubiquitous lipid components observed in virtually all life forms. They are major components of the cell membrane of almost all organisms; and are synthesized, transported, stored, and catabolized by several of them at the same time that they are precursors for the synthesis of more complex lipids and metabolites. Among C18 fatty acids, 18:1 are probably the most abundant in the Earth. The most widely distributed of the 18:1 fatty acids is undoubtedly oleic acid, which plays a key role in unprecedented large number of structure to function life processes. Despite it is impossible to precise the exact moment of the life's evolution when oleic acid appeared among the fatty acids plethora, its crucial role in biological systems can be easily evidenced from the available data and next few years will continue to bring exciting data in this field.

References

[1] H. Kamminga, *Origins Life Evol. Biosph.* 18, 1 (1988).
[2] L. N. Vauquelin and P. J. Robiquet, *Ann. Chim. (Phys.)* 57, 88 (1806).
[3] M. E. Chevreul, *Annalen der Physik* 43, 148 (1813).
[4] F. Miescher, Letters I, IV, V, XVI, and XVII; to Wilhelm His; Tübingen, February 26th (1869).
[5] K. Adamala and P. L. Luisi, *Results Probl. Cell Differ.* 53, 1 (2011).
[6] P. L. Luisi, P. Walde and T. Oberholzer, *Curr. Opin. Coll. Int. Sci.* 4, 33 (1999).
[7] H. Fellermann, *Principles of Evolution*, The Frontiers Collections, Springer (2011).
[8] S. E. Maurer and P-A. Monnard, *Entropy* 13, 466 (2011).
[9] M. M. Haczyc and J. W. Szostak, *Curr. Opin. Chem. Biol.* 8, 660 (2004).
[10] IUPAC-IUB Commission on Biochemical Nomenclature, *Biochem. J.* 171, 21 (1978).
[11] M. T, Nakamura and T. Y. Nara, *Ann. Rev. Nutr.* 24, 345 (2004).
[12] T. J. Brock, J. Browse and J. L. Watts, *Genetics* 176, 865 (2007).
[13] C. R. Noller and R. A. Bannerot, *J. Am. Chem. Soc.* 56, 1563 (1934).
[14] J. K. G. Kramer, C. Cruz-Hernandez, Z. Deng, J. Zhou, G. Jahreis and M. E. R. Dugan, *Am. J. Clin. Nutr.* 79, 1137S (2004).

[15] S. K. Picataggio, N. S. Yadav and Q. Q. Zhu, *United States Patent* No. 20070207528, Mar. 5, 2007.
[16] F. Kaneko, J. Yano and K. Sato, *Curr. Opin. Struct. Biol.* 8, 417 (1998).
[17] F. Kaneko, K. Yamazaki, K. Kitagawa, T. Kikyo and M. Kobayashi, *J. Phys. Chem.* 101, 1803 (1997).
[18] G. J. van der Vusse, M. van Bielsen, J. F. C. Glatz, D. M. Hasselbaink and J. J. F. P. Luiken, *Mol. Cell. Biochem.* 239, 9 (2002).
[19] S. Curry, H. Mandelkow, P. Brick and N. Franks, *Nat. Struct. Biol.* 5, 827 (1998).
[20] S. Lev, *Nat. Rev. Mol. Cell. Biol.* 11, 739 (2010).
[21] J. H. Veerkamp, R. A. Peeters and R. G. Maatman, *Biochim. Biophys. Acta* 1081, 1 (1991).
[22] P. Reis, K. Holmberg, H. Watzke, M. E. Leser and R. Miller, *Adv. Coll. Int. Sci.* 147-148, 237 (2009).
[23] J. J. Kabara, D. M. Swieczkowski, A. J. Conley and J. P. Truant, *Antimicrob. Agents Chemother.* 2, 23 (1972).
[24] D. Walters, L. Raynor, A. Mitchell, R. Walker and K. Walker, *Mycopathologia* 157, 87 (2004).
[25] M. F. Cury-Boaventura, C. Pompéis and R. Curi, *Nutrition* 21, 395 (2005).
[26] M. F. Cury-Boaventura, C. Pompéia and R. Curi, *Clin. Nutr.* 23, 721 (2004).
[27] M. Iwig, D. Glaesser, U. Fass and H. G. Struck, *Exp. Eye Res.* 79, 689 (2004).
[28] J. A. Menéndez, M. D. M. Barbacid, S. Montero, E. Sevilla, E. Escrich, M. Solanas, H. Cortés-Funes and R. Colomer, *Eur. J. Cancer* 37, 402 (2001).
[29] J. A. Menéndez, L. Vellon, R. Colomer and R. Lupu, *Ann. Oncol.* 16, 359 (2005).
[30] M. A. Carluccio, M. Massaro, C. Bonfrate, L. Siculella, M. Maffia, G. Nicolardi, A. Distante, C. Storelli and R. Caterina, *Arterioscler. Thromb. Vasc. Biol.* 19, 220 (1999).
[31] S. Obici, Z. Feng, K. Morgan, D. Stein, G. Karkanias and L. Rosseti, *Diabetes* 51, 271 (2002).
[32] Y. Kimura, *J. Nutr.* 132, 2069 (2002).
[33] S-C. Hsu and C-J. Huang, *J. Nutr.* 136, 1779 (2006).
[34] C. R. Cardoso, S. Favoreto Jr, L. L. Oliveira, J. O. Vancim, G. B. Barban, D. B. Ferraz and J. S. Silva, *Immunobiol.* 216, 409 (2011).
[35] P. C. Calder, *Nutr. Rev.* 56, S70 (1998).
[36] L. R. R. de Araújo, C. F. Scofield, N. M. R. Pastura and W. A. Gonzales, *Mat. Res.* 9, 2 (2006).
[37] R. G. Strickley, *Pharm. Res.* 21, 201 (2004).
[38] P. D. Hughett, *United States Patent* No. 4,350,605, Sep. 21, 1982.
[39] M. Vargas, A. Albors, A. Chiralt and C. González-Martínez, *Postharvest Biol. Technol.* 41, 164 (2006).
[40] G. Cermak and G. Pfeiffer, *United States Patent* No. 4,080,160, Mar. 21, 1978.
[41] S. Statham and R. J. Nelson, *United States Patent* No. 7,902,216, Mar. 8, 2011.
[42] F. Ward, *United States Patent* No. 654,862, Aug. 2, 1977.
[43] L. Rappoport, *United States Patent* No. 5,997,953, Dec. 7, 1999.
[44] T. Bellini, M. Buscaglia, A. Soranno and G. Zanchetta, *Euresis J.* 2, 113 (2010).
[45] A. Chen, R. W. Roberts and J. W. Szostak, *Science* 305, 1474 (2004).
[46] I. Oparin, *The Origin of Life*, Macmillan Co, New York, 1938.

[47] M. Porcar, A. Danchin, V. de Lorenzo, V. A. dos Santos, N. Krasnogor, S. Rasmussen and A. Moya, *Syst. Synth. Biol.* 5, 1 (2011).

[48] T. F. Zhu and J. W. Szostak, *J. Am. Chem, Soc.* 131, 5705 (2009).

[49] S. Ratnasingham and P. D. N. Hebert, *Mol. Ecol. Notes* 7, 355 (2007).

[50] J. Wang, W. F. Chen and Q. X. Li, *Anal. Chim. Acta* 716, 133 (2012).

[51] K. Y. Cho and M. R. J. Salton, *Biochim. Biophys. Acta* 116, 73 (1966).

[52] MIDI Inc., *Sherlock Microbial Identification System References* (2001).

[53] M. Sud, E. Fahy, D. Cotter, A. Brown, E. A. Dennis, C. K. Glass, A. H. Merrill Jr, R. C. Murphy, C. R. H. Raetz, D. W. Russel and S. Subramaniam, *Nucleic Acid Res.* 35, D527 (2007).

[54] K. Aitzetmüller, B. Matthäus and H. Friedrich, *Eur. J. Lipid Sci. Technol.* 105, 92 (2003).

[55] RMIT Lipid Research Group, *Fatty Acid Composition Database* (2001).

[56] B. S. Magalhães, J. Fiamoncini, F. C. Deschamps, R. Curi and L.P. Silva, *Comp. Biochem. Phys. B* 155, 132 (2010).

[57] J. L. Guil-Guerrero, F. F. García Maroto and A. Giménez Giménez, *J. A. Oil. Chem. Soc.* 78, 677 (2001).

[58] Ø. Hammer, D. A. T. Harper and P. D. Ryan, *Palaeontologia Electronica* 4, 1 (2001).

[59] H. Konishi, W. E. Neff and T. L. Mounts, *J. Chromatogr.* 629, 237 (1993).

[60] K. Ichihara and Y. Fukubayashi, *J. Lip. Res.* 51, 635 (2010).

[61] T. Řezanka, *J. Chromatogr. A* 727, 147 (1996).

[62] G. Huys, P. Kämpfer, M. Vancanneyt, R. Coopman, P. Janssen and K. Kersters, *J. Microbiol. Meth.* 28, 89 (1997).

[63] K. J. Ayre and A. J. Hulbert, *J. Nutr.* 126, 653 (1996).

[64] E. Øie, T. Ueland, C. P. Dahl, P. Bohov, C. Berge, A. Yndestad, L. Gullestad, P. Aukrust, R. K. Berge, *J. Intern. Med.* 270, 263 (2011).

[65] Z. Cohen, A. Vonshak and A. Richmond, *J. Phycol.* 24, 328 (1988).

[66] C. Woese, O. Kandler and M. Wheelis, *Proc. Natl. Acad. Sci. USA* 87, 4576 (1990).

[67] C. Mora, D. P. Tittensor, S. Adl, A. G. B. Simpson, B. Worm, *PLOS Biology* 9, 1 (2011).

[68] R. Li and M. M. Watanabe, *Curr. Microbiol.* 49, 376 (2004).

[69] K. Sharathchandra and M. Rajashekhar, *J. Algal. Biomass. Utln.* 2, 2, 83 (2011).

[70] R. Caudales, J. M. Wells and J. E. Butterfield, *Int. J. Syst. Evol. Microbiol.* 50, 1029 (2000).

[71] T. Rezanka, I. Dor, A. Prell and V. M. Dembitsky, *Folia Microbiol.* 48, 1, 71 (2003).

[72] P. Maslova, E. A. Mouradyan, S. S. Lapina, G. L. Klyachko-Gurvich and D. A. Los, *Russ. J. Plant Physiol.* 51, 3, 353 (2004).

[73] M. I. Daneshvar, M. P. Douglas and R. S. Weyantj, *Clin. Microbiol.* 39, 11, 4160 (2001).

[74] J. Cho and S. J. GiovannoniInt., *J. Sys. Evol. Microbiol.* 53, 1031 (2003).

[75] C. Liu and Z. Shao, *J. Sys. Evol. Microbiol.* 55, 1181 (2005).

[76] M. D. Kiran, J. S. S. Prakash, S. Annapoorni, S. Dube, T. Kusano, H. Okuyama, N. Murata and S. Shivaji, *Extremophiles* 8, 401 (2004).

[77] L. E. Nielsen, D. R. Kadavy, S. Rajagopal, R. A. Drijber and K. Nickerson, *Appl. Environ. Microbiol.* 5171 (2005).

[78] K. Y. Cho, M. R. Salton, *Biochim. Biophys. Acta.* 2, 84, 773 (1964).

[79] S. Bowie, M. R. Grigor, G. G. Dunckley, M. W. Loutit and J. S. Loutit, *Soil Bid. Biochem.* 4, 397 (1972).
[80] Y. Chen, X. Cui, R. Pukall, H. Li, Y. Yang, L. Xu, M. Wen, Q. Peng and C. Jiang, *Int. J. Syst. Evol. Microbiol.* 57, 2327 (2007).
[81] S. Makino, H. M. Jenkin, H. M. Yu, And D. Townsend, *J. Bacteriol.* 103, 162, (1970).
[82] E. Kondo and N. Ueta, *J. Bacteriol.* 110, 2, 459 (1972).
[83] Mira, R. Pushker, B. A. Legault, D. Moreira and F. Rodríguez-Valera, *BMC Evol. Biol.* 4, 50 (2004).
[84] R. S. Gupta, T. Mukhtar and B. Singh, *Mol. Microbiol.* 32, 893 (1999).
[85] C. Woese and G. Fox, *Proc. Natl. Acad. Sci. USA* 74, 5088 (1977).
[86] J. Ding, R. Zhang, Y. Yu, D. Jin, C. Liang, Y. Yi, W. Zhu and J. Xia, *Braz. J. Microbiol.* 42, 514 (2011).
[87] J. P. Bowman, J. H. Skerratt, P. D. Nichols and L. I. Sly, *Microbiol. Ecol.* 85, 15 (1991).
[88] J. P. Bowman, L. I. Sly, P. D. Nichols and A.C. Hayward, *Int. J. Syst. Bacteriol.* 43, 735 (1993).
[89] J. P. Bowman, S. A. McCammon, J. H. Skerratt., *Microbiology-UK* 143, 1451 (1997).
[90] S. N. Dedysh, W. Liesack, V. N. Khmelenina, N. E. Suzina, Y. A. Trotsenko, J. D. Semrau, et al. *Int. J. Syst. Evol. Microbiol.* 50, 955 (2000).
[91] S. N. Dedysh, V. N. Khmelenina, N. E. Suzina, Y. A. Trotsenko, J. D. Semrau, W. Liesack et al. *Int. J. Syst. Evol. Microbiol.* 52, 251 (2002).
[92] Dedysh SN, Berestovskaya YY, Vasylieva LV, Belova SE, Khmelenina VN, Suzina NE et al. *Int. J. Syst. Evol. Microbiol.* 54, 151 (2004).
[93] P. F. Dunfield, V. N. Khmelenina, N. E. Suzina, Y. A. Trotsenko and S. N. Dedysh, *Int. J. Syst. Evol. Microbiol.* 53, 1231 (2003).
[94] P. F Dunfield, A. Yuryev, P. Senin, A. V. Smirnova, M. B. Stott, S. B. Hou et al. *Nature* 450, 879 (2007).
[95] B. T Eshinimaev, K. A Medvedkova, V. N Khmelenina, N. E Suzina, G. A. Osipov, A. M Lysenko et al. *Microbiology* 73, 448 (2004).
[96] J. Heyer, U. Berger, M. Hardt and P. F. Dunfield, *Int. J. Syst. Evol. Microbiol.* 1817 (2005).
[97] M. G. Kalyuzhnaya, S. M. Stolyar, A. J. Auman, J. C. Lara, M. E. Lidstrom and L. Chistoserdova, *Int. J. Syst. Evol. Microbiol.* 2345 (2005).
[98] C. Knief, A. Lipski and P. F. Dunfield, *Appl. Environ. Microbiol.* 69, 6703 (2003).
[99] J. Tsubota, B. T. Eshinimaev, V. N. Khmelenina and Y. A. Trotsenko, *Int. J. Syst. Evol. Microbiol.* 1877 (2005).
[100] Wartiainen, A. G. Hestnes, I. R. McDonald and M. M. Svenning, *Int. J. Syst. Evol. Microbiol.* 56, 109 (2006).
[101] Wartiainen, A. G. Hestnes, I. R. McDonald and M. M. Svenning, *Int. J. Syst. Evol. Microbiol*, 56, 541 (2006).
[102] M. G. Wise, J. V. McArthur and L. J. Shimkets, *Int. J. Syst. Evol. Microbiol.* 51, 611 (2001).
[103] M. Vessal, H. R. Rmxr and P. Pakzad, *Exper. Parasitol.* 36, 455 (1974).
[104] W. Leon, A. M. Montelro, C. S. Alwano, M. J. G. Esteves and J. Angluster, *Acta Tropica* 46, 131 (1989).
[105] G. A. Thompson Jr. and Y. Nozawa, *Annu. Rev. Microbiol.*, 26, 249 (1972).

[106] Véra, C. Desvilettes, A. Bec and G. BourdierAquat, *Microb. Ecol.* 25, 271 (2001).
[107] V. Nakamura, L. Waldow, S. R. Pelegrinello, T. Ueda-Nakamura, B. A. Abreu Filho and B. P. Dias Filho, *Mem. Inst. Oswaldo Cruz* 94, 4, 519 (1999).
[108] J. K. Volkman, S. W. Jeffrey, P. D. Nichols, G. I. Rogers and C. D. Garland, *J. Exp. Mar. Biol. Ecol.* 3, 219 (1989).
[109] N. A. Dijkman, H. T. S. Boschker, L. J. Stal and J. C. Kromkamp, *J. Sea Res.* 63, 62 (2010).
[110] F. Goecke, V. Hernández, M. Bittner, M. González, J. Becerra and M. Silva, *Rev. Biol. Mar. Oceanog.* 45, 2, 325 (2010).
[111] X. Zhao, H. Wang, J. You and Y. Suo, *Chromatographia* 66, 197 (2007).
[112] M. Graeve, G. Kattner, C. Wiencke and U. Karsten, *Mar. Ecol-Prog. Ser.* 231, 67 (2002).
[113] X. Bao, S. Katz, M. Pollard and J. Ohlrogge, *Proc. Natl. Acad. Sci. USA* 99, 7172 (2002).
[114] R. Przybylski, *J. Archaeol. Sci.* 26, 83 (1999).
[115] S. Mongrand, A. Badoc, B. Patouille, C. Lacomblez, M. Chavent and J. Bessoule, *Phytochemistry* 66, 549 (2005).
[116] S. Mongrand, A. Badoc, B. Patouille, C. Lacomblez, M. Chavent, C. Cassagne and J. Bessoule, *Phytochemistry* 58, 101 (2001).
[117] R. Caudales, J. M. Wells, A. D. Antoine and J. E. Butterfield, *Int. J. Syst. Bact.* 45, 364 (1995).
[118] M. I. Gurr, *Lipids* 6, 266 (1971).
[119] G-F. Wu, Z-J. Feng, W-L. Ma, X-J. Zhou, K-C. Lang, *Botany*. Beijing: The High Education Press House (1990).
[120] J. D. Weete, M. S. Fuller, M. Q. Huang, and S. Gandhi, *Exp. Mycol.* 13, 183, (1989).
[121] L. Karlinski, S. Ravnskov, B. Kieliszewska-Rokicka and J. Larsen, *Soil. Biol. Biochem.* 39, 854 (2007).
[122] P. D. Stahl and M. J. Klug, *Appl. Environ. Microbiol.* 62, 11, 4136 (1996).
[123] L. Barros, P. Baptista, D. M. Correia, Susana Casal, B. Oliveira and I. C. F. R. Ferreira, *Food Chem.* 105, 140 (2007).
[124] Tylicki, M. Siemieniuk, P. Dobrzyn, G. Ziolkowska, M. Nowik, U. Czyzewska and A. Pyrkowska, *Mycoses* 55, 106 (2012).
[125] J. Guarro, J. Gené and A. M. Stchigel, *Clin. Microbiol. Rev.* 12, 454 (1999).
[126] D. S. Hibbett et al. *Mycol. Res.* 111, 509 (2007).
[127] Schreiber, G. Wörheide and V. Thiel, *Chem. Phys. Lipids* 143, 29 (2006).
[128] N. M. Carballeira, F. Shalabi, C. Cruz, J. Rodriguez, and E. Rodriguez, *Comp. Biochem. Physiol. B* 100, 3, 489 (1991).
[129] H. Yamashiro, H. Oku, H. Higa, I. Chinen and K. Sakai, *Comp. Biochem. Physiol. B* 122, 397 (1999).
[130] J. W. Moore, *Estuar. Coast Mar. Sci.* 4, 215 (1976).
[131] F. Culkin, R. J. Morris, *Deep-Sea Res.* 16, 109 (1969).
[132] P. Ward, R. S. Shreeve and G. Gcripps, *J. Plankton Res.* 18, 8, 1439 (1996).
[133] J. M. Guerra-García, I. Martínez-Pita and M. L. Pita, *Sci. Mar.* 68, 4, 501 (2004).
[134] S. N. Thompson, *Comp. Biochem. Physiol. B* 45, 467 (1973).
[135] M. Yahyavi, M. Afkhami, A. Javadi, M. Ehsanpour, A. Khazaali, R. khoshnood and A. Mokhlesi, *Afr. J. Biotechnol.* 11, 12, 2862 (2012).

[136] R. Chedoloh, T. T. Karrila and P. Pakdeechanuan, *Int. Food. Res. J.* 18, 783 (2011).
[137] T. Fukushima, K. Abe, A. Nakagawa and Y. Yamane, *Int. J. Parasitol.* 18, 27 (1988).
[138] Y. Özogul, F. Özogul, E. Ççek, A. Polat and E. Kuley, *Int. J. Food. Sci. Nutr.* 60, 464 (2009).
[139] H. Hornung, A. Sukenik and G. P. Gabrielides, *Mar Pollut. Bull*, 28, 448 (1994).
[140] T. A. Kozlova and S. V. Khotimchenko, *Comp. Biochem. Phys. B* 126, 477 (2000).
[141] M. Kaneniwa, S. Miao, C. Yuan, H. Iida and Y. Fukuda, *J. Am. Oil. Chem. Soc.* 77, 825 (2000).
[142] R. Tanakol, Z. Yazıcı, E. Sener and E. Sencer, *Lipids* 34, 3, 291 (1999).
[143] R. Chedoloh, T. T. Karrila and P. Pakdeechanuan, *Int. Food Res. J.* 18, 783 (2011).
[144] B. S. Magalhães, J. Fiamoncini, F.C. Deschamps, R. Curi and L.P. Silva, *Comp. Biochem. Phys. B* 155, 132, (2010).
[145] L. E. Gutierrez and R. C. M. Da Silva, *Sci. Agric.* 50, 478 (1993).
[146] Y. Özogul, F. Özogul and S. Alagoz, *Food Chem.*, 103, 217 (2007).
[147] E. Méndez, R. M. González, G. Inocente, H. G.Iudice and M. A. Grompone, *J. Food Comp. Anal.* 9, 163 (1996).
[148] M. M. Ramos Filho, M. I. L. Ramos, P. A. Hiane and E. M. T. Souza, *Ciênc. Tecnol. Aliment.* 28, 361 (2008).
[149] T. Kozlova and S. Khotimchenko, *Comp. Biochem. Physiol.* 105, 97 (1993).
[150] G. Aggelousis and E. S. Lazos, *J. Food Comp. Anal.* 4, 68 (1991).
[151] D. S. Nichols, D. Williams, G. A. Dunstan, P. D. Nichols and J. K. Volkman, *Comp. Biochem. Physiol. B* 107, 357, (1994).
[152] E. Méndez, J. Sanhueza, S. Nicto, H. Speisky and A. Valenzuela, *J. Am. Oil. Chem. Soc.* 75, 79 (1998).
[153] M. D. McCue, *Comp. Biochem. Phys. B* 151, 239 (2008).
[154] R. G. Ackman, E. J. Macpherson and R. K. O'dorcomp, *Comp. Biochem. Physiol. B* 98, 423 (1991).
[155] T. Fukushima, K. Abe, A. Nakagawa and Y. Yamane, *Int. J. Parasitol.* 18, 27 (1988).
[156] G. F. Mcmullin, S. C. Smith and P. A. Wright, *Comp. Biochem. Physiol.* 26, 211 (1968).
[157] L. D. Caldwell, *Comp. Biochem. Physiol. B*, 44, 493 (1973).
[158] P. F. Romanelli, E. Trabuco, A. B. Scriboni, J. V. Visentainer and N. E. Souza, *Organo Oficial de la Sociedad Latinoamericana de Nutrición* 58, 2, 201 (2008).
[159] J. Shen, R. Li and F. Gao, *Comp. Biochem. Phys. B* 142, 293 (2005).
[160] R. Przybylski, *J. Archaeol. Sci.* 26, 83 (1999).
[161] J. D. Wood, M. Enser, A. V. Fisher, G. R. Nute, P. R. Sheard, R. I. Richardson, S. I. Hughes and F. M. Whittington, *Meat Sci.* 78, 343 (2008).
[162] F. Shahidi and Y. Zhong, Bailey's Industrial Oil and Fat Products, Sixth Edition (2005).
[163] G. W. Thiemann, S.J. Iverson and I. Stirling, *Mar Mammal Sci* 24, 91 (2008).
[164] O. Mahgoub, A.J. Khan, R. S. Al-Maqbaly, J.N. Al-Sabahi, K. Annamalai and N. M. Al-Sakry, *Meat Sci.* 61, 381 (2002).
[165] R. Guitart, A. M. Silvestre, X. Guerrero and R. Mateo, *Comp. Biochem. Phys. B,* 124, 439 (1999).
[166] G. F. Mcmullin, S. C. Smith and P. A. Wright, *Comp. Biochem. Physiol.* 26, 211 (1968).

[167] F. Paganelli, J. M. Maixent, M. J. Duran, R. Parhigar, G. Pieroni and S. Sennoune, *Int. J. Cardiol.* 78, 27 (2001).

[168] R. Plasenzotti, U. Windberger, F. Ulberth, W. Osterode and U. Losert, *Clin. Hemorheol. Microbiol.* 37, 237 (2007).

[169] S. P. Makarenko, Y. M. Konstantinov, S. V. Khotimchenko, T. A. Konenkina and A. Sh. Arziev, *Russ. J. Plant. Physiol.* 50, 4, 487 (2003).

[170] H. A. Schwertner and J. B. Biale, *J. Lipid Res.* 14, 235 (1973).

[171] F. T. Wolf, J. G. Coniglio and J. T. Davis, *Plant. Physiol.* 83 (1961).

[172] R. P. Poincelot, *Plant. Physiol.* 58, 595 (1976).

[173] W. Fiehn, J. B. Peter, J. F. Mead and M. Gap-Elepano, *J. Biol. Chem.* 246, 18, 5617 (1971).

[174] T. Williams and I. P. Thompson, *Arch. Virol.* 140, 975 (1995).

[175] H. A. Blough, *J. Gen. Virol.* 12, 317 (1971).

In: Oleic Acid: Dietary Sources, Functions and Health Benefits ISBN: 978-1-62618-332-2
Editor: Luciano Paulino Silva

Chapter 2

OLEIC ACID: NATURAL VARIATION AND POTENTIAL ENHANCEMENT IN OILSEED CROPS

Noelle A. Barkley and Ming Li Wang*
USDA-ARS, PGRCU, Griffin, Georgia, US

ABSTRACT

Oleic acid is a monounsaturated omega 9 fatty acid (MUFA, C18:1) which can be found in various plant lipids and animal fats. Unlike omega 3 (linolenic acid, C18:3) and omega 6 (linoleic acid, C18:2) fatty acids which are essential because they cannot be synthesized by humans and must be obtained from dietary sources, omega 9 is not an essential fatty acid and can be synthesized de novo.

Even though oleic acid is not essential, it plays important biological roles in both plants and animals. Consumption of high oleic oils leads to a wide range of beneficial effects on promoting human health (from enhancing immune responses to reducing risks of cardiovascular diseases). Plant oils are one of the most important dietary sources of oleic acid.

The level of oleic acid in plant oils is one of the key factors to determine their nutritional quality and total shelf life. Significant variability in the levels of oleic acid has been identified in different plant oils. Natural variation of oleic acid levels also exists in the oil extracted from the same species likely due to genotype, quantitative trait loci, or environmental effects.

Plant breeders have been using traditional breeding methods to enhance the level of oleic acid in oilseed crops, but the level of enhancement is limited. Genetic modification (including chemical mutagenesis, physical radiation, random bombardment, target-specific modification) has played a big role in altering the total fatty acid composition and has been successfully used on oilseed crops (including soybean, canola, sunflower, cotton, and peanut) for enhancing oleic acid levels. Here in this chapter, we review the known biological function of oleic acid, the beneficial effects on human health, dietary sources, and natural variation and potential enhancement on the level of oleic acid in crop plants.

* E-mail address: MingLi.Wang@ars.usda.gov.

INTRODUCTION

Fatty acids are chemically characterized as carboxylic acids with long hydrocarbon chains which vary in length from 4-28. There are two main types of fatty acids termed saturated and unsaturated which can further be classified as monounsaturated and polyunsaturated. Saturated fatty acids have no double bonds between carbon atoms; whereas, unsaturated fatty acids have at least one carbon double bond (monounsaturated) or may have several double bonds (polyunsaturated). Fatty acids either react with glycerol to form triacylglyceride or when not attached to other molecules are known as free fatty acids.

Triacylglycerides are energy storage reserves which provide more energy than carbohydrates. Fatty acids are important sources of energy to all living organisms because when metabolized ATP is produced which provides vital energy needed within cells.

Oleic acid (C18:1 cis-9), a monounsaturated omega-9 fatty acid is one of the most common fatty acids found in plants and animals. This fatty acid is a major monounsaturated fat of membrane glycerolipids [1]. Chemically characterized as pale yellow to brownish-yellow oily liquid with a lardlike odor, it is insoluble in water, has a molar mass of 282.46 g/mol, a high boiling point of 360°C, and a melting point of 16.3°C [2]. The chemical formula for oleic acid is expressed as $CH_3(CH_2)_7CH{=}CH(CH_2)_7COOH$ or $C_{18}H_{34}O_2$ with a CAS registry number of 2027-47-6. The IUPAC name for oleic acid is cis-9-octadecenoic acid. Oleic acid can be distinguished from other 18 carbon fatty acids by a single double bond in the middle of the chain of 18 hydrocarbons.

Plants synthesize a large variety of fatty acids; however, only a few are common constituents [3] in a cell. Synthesis of oleic acid occurs *de novo* in the plastids of plant leaves and developing seeds which can then either enter the prokaryotic pathway glycerolipid synthesis in the inner envelope of the chloroplasts or can be exported from the plastid as CoA thioesters to enter the eukaryotic glycerolipid synthesis pathway [1, 4, 5, 6]. Production of oleic acid is dependent on synthesis of stearic acid (18:0) in the cell.

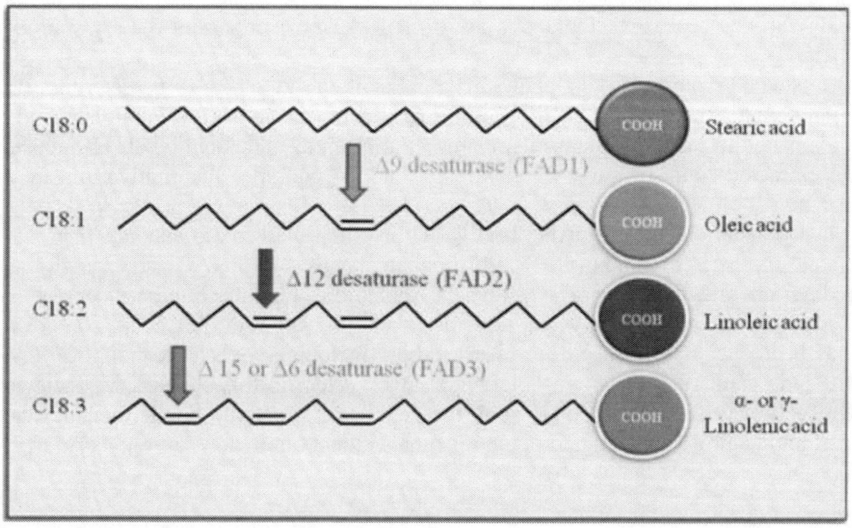

Figure 1. Chemical structure of 18 carbon fatty acids and the desaturase genes which convert these fatty acids by adding double bond.

The stearoyl-acyl carrier protein (ACP) or Δ-9 desaturase present in the stroma of the chloroplast converts stearic into oleic by adding a carbon double bond into the hydrocarbon chain. In the fatty acid biosynthetic pathway, oleic acid can be further converted to a polyunsaturated fat, linoleic acid (C18:2), by desaturation (adding another double bond at position ω-6) or converted into a saturated fat, stearic acid (C18:0) by hydrogenation of the double bond (Figure 1). Conversion of oleic to stearic acid has been demonstrated to occur from microbial biohydrogenation [7].

GENERAL BIOLOGICAL FUNCTIONS OF OLEIC ACID

Fatty acids found in plants are the principal components of membrane phospholipids and triacylglycerol (TAG) storage [8]. Fatty acids are typically bound to phospholipids and cholesterol esters [9]. Since biological membranes are composed of fatty acids, the membrane composition can be influenced by the type of fat consumed in a diet; therefore, diets rich in oleic acid (such as a Mediterranean diet) have been associated with increases of oleic acid in the plasma membrane and modulations in the function of various proteins [9]. Changes in oleic acid membrane levels could affect the function of proteins.

Plants utilize unsaturated fatty acids to modulate membrane fluidity ultimately helping to protect the plants' survival. The ability to adjust unsaturated fatty acids levels in biological membranes by modulating fatty acid desaturase genes allow plants to resist biotic and abiotic stresses that occur in their environment [10]. For example, one study showed that contaminated heavy metal soils produced significant changes in the fatty acid composition [increased oleic and linoleic (C18:2), decreased linolenic (C18:3)] in the leaves of tomato plants; however, the fatty acid composition in the roots and cotyledons were unchanged compared to a control [11]. Fatty acids are also involved in regulating the colonization of seed pathogens by controlling fungal development and mycotoxin synthesis [10]. Free oleic acid levels in chloroplast regulate programmed cell death, and systematic acquired resistance [10]. Further, in plants changes in oleic acid levels results in fluctuations of phytohormones such as salicylic acid and jasmonic acid which play a role in pathogen defense response [1]. Reductions of oleic acid levels in *Arabidopsis* caused altered defense signaling (upregulated expression of salicylic acid mediated pathway and downregulated jasmonic acid mediated pathway) which were shown to be rescued by elevating endogenous 18:1 levels [1, 5]. Low oleic acid levels have also been shown to induce similar defense responses in other plants such as soybean (*Glycine max*) and rice (*Oryza sativa*) [12]. Therefore, overall research has demonstrated that abiotic and biotic stresses cause plants to modify their fatty acid composition in response to different types of stresses.

Fatty acids not only function in helping a plant to modulate certain biotic and abiotic stress, they also play an important role in storage and flavor of oilseed crops. The fatty acid composition in oilseed crops is a major determinant of their overall quality [13, 14]. Fatty acids influence seed quality due to the likelihood of oxidation of the oil which can break down over time and ultimately produce noxious odors or off flavors in stored food products. Consequently, the shelf life of oilseed products is time limited. Generally, monounsaturated fatty acids are less prone to oxidation during processing and storage than unsaturated fatty acids [13, 15] because the double bonds in polyunsaturated fats break down to produce acids, aldehydes, ketones, and hydrocarbons [13]. Therefore, oilseeds that contain a high amount of oleic acid (monounsaturated fat) and low amounts of polyunsaturated fats are generally preferred by manufacturers and consumers. Consequently, a concerted effort has been placed on identifying natural genetic variation in oilseeds followed by conventional breeding or employing molecular biotechnology techniques to develop oilseed crops which contain high levels of oleic acid and low levels of linoleic (C18:2) or linolenic (C18:3) acid.

HEALTH EFFECTS AND DIETARY SOURCES OF OLEIC ACID

There are many choices of vegetable oils or solid fats for consumption in the human diet. Typically, vegetable oils are liquid at room temperature due to the relatively high content of unsaturated fatty acids. Fats, however, are typically solid because they are mainly composed of saturated fatty acids. Vegetable oils can be converted from a liquid to a solid or semi-solid through a process known as hydrogenation which alters the physical properties of the oil. Catalytic hydrogenation of edible fats was patented in 1903 and by the middle of the 20th century millions of tons of oil was hydrogenated or partially hydrogenated [16]. This process reduces or removes the double bonds in the hydrocarbon chains and thereby increases the melting point of the oil [17]. Hydrogenation not only changes a liquid fat into a solid or semi-solid, but it also significantly increases the shelf life by reducing less stable polyunsaturated fatty acids that tend to go rancid via oxidation into more stable, longer lasting fatty acids. However, the problem with this chemical processing is that additional side reactions occur during hydrogenation converting naturally occurring *cis* double bonds to the *trans* configuration [17]. Trans fats produced via hydrogenation have been associated with increasing LDL cholesterol, decreasing HDL cholesterol, increasing total cholesterol content, and an increasing risk of coronary heart disease [16, 18]. Cooking oils with high amounts of oleic acid may limit the need for catalytic hydrogenation or deem it unnecessary due to the stability and long shelf life that these oils bestow, since oils with a large amount of monounsaturated fats are far less prone to oxidation.

Some dietary sources of oleic acid are olive oil, soybean oil, canola oil, sunflower oil, safflower oil, peanut oil, sesame oil, almonds, avocados, cashews, hazelnuts, macadamia nuts, pecans, pistachio, peanuts, and peanut butter. Oleic acid is one of the predominant fatty acids in some animal foods (poultry and pork) in Western diets [19]. Vegetables, however, are the main sources of oleic (18:1), linoleic (18:2), and linolenic (18:3) fatty acids [6]. The Mediterranean diet characterized by a large intake of fat (mainly monounsaturated fatty acids such as oleic and limited amount of saturated fats) has been coupled with low chronic disease and long life expectancies in human populations [20]. Oleic acid has been associated with a reduction in systolic blood pressure [21], reducing triacylglycerol [22], guarding low density lipoproteins (LDL) from oxidative modification [23], reduction of LDL cholesterol levels [24], helping maintain good cholesterol levels known as high density lipoproteins (HDL), reducing blood glucose levels in type II diabetes [25], reducing the risk of breast cancer [26], and slowing down atherosclerosis [27]. Oleic acid synthetic derivatives have also been developed as anticancer and antihypertensive drugs due to their effects on membrane structure and signaling [9]. Therefore, consuming a diet rich in oleic acid can have a significant impact on human health.

NATURAL VARIATION AND POTENTIAL ENHANCEMENT ON THE LEVEL OF OLEIC ACID IN OILSEED CROPS

Soybean [*Glycine max* (L.) Marr.], canola (*Brassica napus* L.), sunflower (*Helianthus annuus* L.), peanut (*Arachis hypogaea* L.), and cotton (*Gossypium hirsutum* L., a major cultivated species) are the five major oilseed crops in the world and the oil production from their seeds were 8.8, 3.3, 2.3, 1.3, and 1.0 billion gallons in the year of 2002/2003, respectively [28]. Natural genetic variation exists on the fatty acid composition and total oil content among these five major species (Table 1). Oilseeds such as rapeseed, safflower, and sunflower contain approximately 40% oil by weight [4]. Generally, peanut seed contain a relatively higher amount of oil (49.2%) than many of the other oilseed crops. In terms of fatty acid composition, canola, and peanut oil contain higher amounts of oleic acid (59.2% and 47.3%) and lower amount of linoleic acid (20.1% and 31.1%) whereas soybean, sunflower, and cottonseed oil contain lower amounts of oleic acid (22.8%, 26.7%, and 15.6%) and higher amounts of linoleic acid (51.0%, 61.2%, and 55.6%). Overall, the nutritional quality of canola and peanut oil is better than soybean, sunflower, and cottonseed oil. These major edible oilseeds have similar fatty acid compositions in which 18 carbon fatty acids with one to three double bonds are the major constituents of the oil composition which indicates that the uses of these major oils are interchangeable since the fatty acid profile are similar [29].

The oils from these five major oilseed crops are not only used for human consumption but they also have other applications such as the production of biodiesel, detergents, and lubricants. Therefore, oilseeds can be exploited because they are low polluting chemical factories that convert energy from sunlight into a variety of valuable chemical structures with a multitude of nonfood uses [17]. Feedstocks used in biodiesel production that are composed mostly of monounsaturated fats (oleic acid) helps balance the tradeoff between cold flow and oxidative stability. High oleic (HO) oil converted to biodiesel had a significant reduction in nitrogen oxides (NO_x) compare to a normal oleic oil biodiesel [30] which helps reduce the amount of pollution compared to normal oleic oil that is converted to biodiesel. Given oilseeds inherent value in cooking and for energy, vegetable oils worldwide consumption has increased by over 50% in the past decade and prices of palm, soybean, and canola have doubled since 2000 [31].

Table 1. Typical fatty acid composition and oil content in traditional varieties from five major oilseed crops

Fatty acid/Species oil	Soybean [a]	Canola [b]	Sunflower [c]	Cottonseed [d]	Peanut [e]
Oleic acid (C18:1)%	22.8	59.2	26.7	15.6	47.3
Linoleic acid (C18:2)%	51.0	20.1	61.2	55.6	31.1
Linolenic acid (C18:3)%	6.8	11.1	0.2	0.3	0.1
Palmitic acid (C16:0)%	10.3	4.0	5.8	23.0	10.7
Stearic acid (C18:0)%	3.8	1.8	3.8	2.3	3.7
Others%	5.3	3.8	2.3	3.2	7.1
Oil content	23.0	43.1	42.2	22.0	49.2

Sources: [a] FAO corporate document repository, National Research Council, and [73]; [b] www.grainscanada.gc.ca; [c] www.sunflowernsa.com and [74]; [d] [75]; and [e] Wang et al. unpublished results (personal data).

Unfortunately, extracted oils do not naturally contain high levels of oleic acid (>70%) among many major cultivated oilseed species [32]. Further, the level of oleic acid produced in the seed can be influenced not only by the genotype, but also by the environment, the temperature, sowing dates, soil types, maternal effects, and irrigation conditions [4, 33, 34, 35]. Nevertheless, traditional plant breeding, marker assisted selection, mutagenesis, transgenes, RNAi, and other molecular biotechnology techniques have allowed the manipulation of fatty acid content in oilseed crops to enhance oleic acid levels from normal levels of 40-60% to 75-85%, as well as, eliciting changes in other desired fatty acids. Much of this work has been accomplished by targeting the fatty acid desaturase gene (*FAD2*) which is responsible for producing an enzyme (microsomal oleoyl-PC desaturase) that removes hydrogen atoms from the hydrocarbon chain creating carbon double bonds thereby converting oleic acid into linoleic acid. There is interest in exploring the degree to which oil composition can be modified in oilseed crops because relatively large alterations in fatty acid composition can be made without deleterious effects on the development of the plant [4]. Examples of improving the oil quality using various techniques will be discussed according to individual oilseed crops.

Soybean

Soybean (*Glycine max*) is the number one cultivated oilseed crop in the United States, Brazil, Argentina, China, and India. Generally, the seeds contain 20% oil [36]. Soybean seed oil mainly contains five major fatty acids: 51% linoleic acid (C18:2), 22.8% oleic acid (C18:1), 10.3% palmitic acid (C16:0), 6.8% linolenic acid (C18:3), and 3.8% stearic acid (C18:0) [Table 1]. The amount of oleic acid only represents about $1/5^{th}$ of the soybean oil. Whether soybean oil is used as edible oil or as a feedstock for biodiesel production, the level of oleic acid needs to be enhanced. Fatty acid composition in soybean has been shown to be influenced by genotype and temperature in that a reduction of 18:1 and an increase of 18:2 and 18:3 occurred in the mature seeds in response to cooler growing temperatures [37]. One of the drawbacks of soybean oil is the low oxidative stability, off-flavors, and rancidity that can occur due to its high levels of polyunsaturated fatty acids; but modification can improve its overall quality [36]. When soybean oil is used as a feedstock for biodiesel production, the amounts of saturated fatty acids such as palmitic acid (16:0) and stearic acid (18:0) need to be decreased for oxidative stability, and the amount of polyunsaturated fatty acids such as linoleic (18:2) acid and linolenic (18:3) acid need to be decreased to improve the cold flow properties of the biodiesel. Further, increasing the oxidative stability of soybean oil by increasing oleic acid content will allow this oil to be used for industrial applications such as spraying oils and machine lubricants [38].

There have been several studies reporting alterations in fatty acid content in soybean using mutagenesis or RNAi technology. X-rays were utilized to irradiate dry soybean seeds to produce a high oleic acid mutant. The mutant, M23, had 46.1% oleic acid, twice as much oleic acid as typically found in the normal untreated soybean seeds [39]. EMS and γ-radiation was used in combination to treat soybean cultivar 'MACS 450'. Treating with two mutagens proved to be highly effective in increasing the variability of fatty acid content mainly by producing lines with higher levels of oleic acid (40%) and lower levels of linolenic acid (4%) [36]. Transgenic soybean lines were developed by using ribozyme terminated antisense and standard antisense constructs with the *FAD2-1* and *FatB* open reading frames (ORF) from soybean to knock down the accumulation of linoleic acid (C18:2) and palmitic acid (C16:0) and thereby increase oleic acid (C18:1) accumulation. This study showed that the transgenic lines varied on oleic acid content from 57-85% which was a significant increase compared to 17.9% in the wild type [40].

Natural genetic variation exists in plant germplasm which can be employed once identified to manipulate fatty acid composition through traditional breeding and selection. Conventional breeding approaches were used to exploit natural variation in oleic acid levels which led to developing mid-oleic acid (~30-70%) genotypes [41]. Recently, spontaneous mutant alleles on *FAD2-1A* and *FAD2-1B* (desaturase genes which convert oleic to linoleic) have been identified in soybean germplasm accessions. When PI 603452 (containing mutant *FAD2-1A* allele) and PI 283327 (containing mutant *FAD2-1B* allele) were used as parents and crossed, progenies with 82-86% oleic acid were identified by molecular marker genotyping assays. Furthermore, the oil and protein content of these high oleic (HO) lines were very similar to traditional varieties [38]. This case demonstrated a good example of non-transgenic HO acid cultivars developed by exploring natural germplasm through traditional breeding methods.

Canola

Canola is the number one oilseed crop in Canada. Canola is a specific edible type of rapeseed in which the seed oil contains a low amount of erucic acid (C22:1), less than 2%. (Erucic acid is a monounsaturated fatty acid produced from elongating the hydrocarbon chain of oleic acid which can be used for surfactants and biodiesel and is prevalent in rapeseed). As cooking oil, canola has a higher smoke point (242°C) than soybean oil (234°C) which makes it better oil for frying because once the smoke point is reached the flavor and nutrition of the oil begins to degrade. Canola typically produce oils composed of approximately 59.2% oleic acid (C18:1), 20.1% linoleic acid (C18:2) and 11.1% linolenic acid (C18:3) [Table 1]. *Brassica juncea* has even lower concentrations of oleic acid at 45% which is significantly lower than either *B. napus* or *B. rapa* [42]. Although canola oil contains a relatively high amount of oleic acid, high levels of linolenic acid (a highly unsaturated fatty acid) in the oil are undesirable because it can easily lead to oxidization causing off-flavors and rancidity [43].

Genetic variation to reduce polyunsaturated fatty acids by traditional breeding and selection in *Brassica* germplasm is limited [44]. However, mutagenesis, RNAi, and transformation have provided a means to develop HO *Brassica* lines. *B. napus* and *B. rapa* (allotetraploid and diploid, respectively) were treated with ethyl methanesulfonate (EMS) which produced lines containing 87% oleic acid [44]. In a separate study, genetic transformation modified the oleic acid content in *B. napus* and *B. juncea* to above 80% and 73%, respectively by cosuppression of the *FAD2* gene [42]. Targeted sequencing of the *FAD2* gene in mutant *B. napus* lines revealed a single nucleotide transition mutation (C to T) which caused a premature stop codon producing a protein of 185 amino acids instead of the wild type protein of 384 amino acids causing a nonfunctional desaturase and accumulation of oleic acid in the mutant line seeds [43]. This functional SNP was developed into a marker (Invader® assay) to be used for marker assisted selection in conjunction with traditional breeding. Recently, HO canola (containing 74% and 78.0% oleic acid) were also developed via gene silencing by RNAi or expression of the gene *BrFAD2-1*in an antisense orientation, respectively [45, 46].

Sunflower

Sunflower (*Helianthus annuus* L.) is one of the most cultivated oilseed crops. Russia is the main producer of sunflower in the world, following by Ukraine and the European Union. This crop is the second largest hybrid crop after maize and fifth largest among oilseeds [47]. Sunflowers can be used in food and non-food industries (such as cooking oil, birdseed, cut flowers, and biodiesel). Further, they have significant variability in traits such as disease, pest resistance, quality and composition of compounds in the seeds; but domestication has reduced 40-50% of this diversity comparatively to the wild germplasm [47]. Sunflower oil naturally contains up to 90% unsaturated fatty acids consisting of oleic (~16-19%) and linoleic (~68-72%) fatty acids [48]. Environmental effects can have an impact on sunflower oil composition. Drought increased oleic acid content in HO lines, but decreased the oleic content in ordinary hybrids [49].

Improving sunflower oil has been achieved by mutagenesis via γ-rays or mutagenic chemicals such as ethyl methanesulfonate (EMS) and dimethyl sulfate (DMS). The first of which was achieved by Soldatov in 1976 who treated VNIIMK 8931 with dimethyl sulfate to produce a HO mutant known as 'Pervenets' [50]. 'Pervenets' has been used worldwide as a source to develop HO sunflower hybrids. Further work with chemical mutagens allowed HO hybrids to be developed with altered tocopherol levels which will provide the oil 10-20 times greater oxidative stability than conventional oil [48]. An EMS mutagenized population generated from a HO (90% C18:1) inbred line GV342 was utilized in a TILLING experiment to assess mutations in *fad2-1* as well as other candidate genes. EMS generated three mutations (missense, silent, and a mutation in the intron) in *fad2-1* with a total mutation frequency of 1/332kb [47]. Recently, a technique on silencing oleate desaturase (OD) gene via siRNA was also developed to enhance oleic acid content in sunflower oil for breeding programs [51].

Safflower

Safflower (*Carthamus spp.*) is an herbaceous annual that is commercially cultivated for the oil extracted from the seeds. Typically, safflower has about 30% oil in normal hull types; but, the thin hull types contain 46% oil [52]. Besides for the oil, safflower is also cultivated for its meal (animal feed), flowers, extracting carthamin (a dye used to color foods, beverages, and cosmetics), and industrial purposes such as preparing varnishes and textile dyes [53]. Variability of fatty acid composition exists among safflower cultivars. Standard safflower contains approximately 7% palmitic, 3% stearic, 18% oleic, and 72% linoleic [53, 54]. HO safflower lines have been developed with oleic acid content between 75-85%; however, high linoleic safflowers (85%) also are available. HO and high linoleic safflower lines have been demonstrated to be temperature stable; but, lines with near equal ratios of oleic and linoleic are not stable because the amount of oleic acid increases in high temperature or the amount of linoleic acid increases in lower temperatures [55]. Genotype during oil formation has a major effect on the production of oleic and linoleic acids, as well as, sowing date, harvest date, drought and moisture during seed development [53, 56]. Severe drought stress reduced total oil content and the level of saturated (palmitic and stearic) fatty acids, as well as, unsaturated (oleic and linoleic) fatty acids in safflower seed oil [56].

Cotton

Cotton (*Gossypium spp.*) is mainly grown for fiber production and cottonseed oil is its by-product. Heavy fiber requirement makes cotton one of the most important oilseed crops in the world. Cotton seeds are not edible without refinement due to gossypol (phenolic aldehyde) in the seed; yet, they do produce about 16% oil by weight with approximately 15-18% of the extracted oil being composed of oleic acid [57] and about 30% saturated and 52% polyunsaturated (linoleic and linolenic) fat. Although cotton is primarily grown for fiber, it has also been widely used as a source of vegetable oil [58]. The level of oleic in normal cotton seeds is low in comparison to other oilseed crops.

There is very limited genetic variation in cottonseed fatty acid composition. Further, classical breeding techniques and mutation induction have not produced large gains in improving fatty acid composition in cotton because of its allotetraploid nature and due to a lack of significant genetic variation in fatty acid composition [58]. However, oleic acid content was effectively increased from 15% to 47% by *Agrobacterium* mediated transformation to suppress the endogenous FAD2 enzyme activity. This strategy demonstrated that creation of transgenic lines with a mutant *fad2* allele from rapeseed produced 50% transgenic lines with a modified oleic acid phenotype [57].

Another study demonstrated that HO cottonseed could be successfully produced by employing hairpin RNA mediated post-transcriptional gene silencing (RNAi) to knock down gene expression of *ghSAD-1* and *ghFAD2-1* which encode stearoyl acyl carrier Δ9-desaturase and oleoyl-phosphatidylcholine ω6-desaturase, respectively. This transgenic modification produced altered fatty acid compositions with oleic acid content up to 78% and stearic acid content as high as 40%. Furthermore, intercrossing a high stearic with a HO T_1 genotype simultaneously down regulated both *ghSAD-1* and *ghFAD2-1* demonstrating the efficacy of stacked hairpin RNA induced posttranslational gene silencing and novel combinations of stearic and oleic acid levels in the progeny carrying both transgenes [58]. Therefore, much of the improvement in cottonseed oil composition has come through transgenic modification.

Sesame

Sesame (*Sesamum indicum* L.) is regarded in the orient as a health food that increases energy and helps prevents aging. This oilseed crop has multiple uses including food products, pharmaceuticals, cosmetics, perfumery, soaps, paints, and a carrier for insecticides [59]. The oil is known to be highly stable compared to other oils believed to be due to its chemical composition in that it naturally contains endogenous antioxidants such as tocopherol and lignans [sesamin and sesamolin] [60] which may prevent the fatty acids from oxidation. Comparatively to other oilseeds, sesame seeds have a relatively high oil and protein content with 50-60% oil and 25% protein in their seeds. The extracted oil from sesame is predominantly (80%) composed of oleic and linoleic acid typically in about equal proportions. However, as characterized in other oilseed crops, sesame fatty acid composition and oil content is influenced by the environment, genotype, and sowing date [59, 61]. Analysis of sesame oil demonstrated that the black seeds contain significantly lower amounts of total lipids than yellow or white sesame [60]. Natural variation of oil content and fatty acid composition was found when evaluating 103 accessions of *S. indicum* germplasm. The oil content varied from 41.3 to 62.7% and the mean oleic acid content and linoleic acid content in the extracted oil was 37.2% and 45.7%, respectively [59]. Evaluation of germplasm collections allows the identification of desired oil characteristics to make informed decisions on selections of desired parents for breeding programs.

Peanut

Cultivated peanuts (*Arachis hypogaea* L.) are important oilseed crops worldwide, because these allotetraploids ($2n = 4x = 40$) typically contain 50% oil in the seed. Although up to 12 fatty acids have been reported in peanuts, only three fatty acids palmitic (C16:0), oleic (C18:1), and linoleic (C18:2) acids comprise the majority (~90%) of the fatty acids found in peanut oil [13, 62]. The fatty acid composition in the oil extracted from peanuts tends to range between 36-80% oleic and 2-43% linoleic fatty acids [62]. In general, high oleate lines are characterized by 74-80% oleate (C18:1) and less than 7% linoleate (C18:2); whereas, normal oleate can range from 30-60% oleic acid and greater than 10% linoleic [13, 63]. O/L (oleic to linoleic) acid ratios of 10 or greater are generally necessary for a line to be considered HO (Dan Gorbet, personal communication). In normal peanuts (oleate ~30-60%), the fatty acid composition can vary dependent on the genotype of the individual, seed maturity, and environmental interactions such as time of year planted, seasonal variation, insect damage, and soil conditions [33, 35, 62].

The first HO peanut identified was F435, thought to be a spontaneous mutant [61] (Norden et al. 1987) and since its discovery the majority of HO peanut cultivars have been derived from F435 or F435 derivatives. The high oleate phenotype in peanut has been previously demonstrated to originate from two recessive mutations (*ahfad2A* and *ahfad2B*). The wild type homoeologous genes (*ahFAD2A* and *ahFAD2B*) encode microsomal oleoyl-PC desaturase also known as Δ^{12} fatty acid desaturase [13, 63]. The open reading frames (ORF) for *ahFAD2A* and *ahFAD2B* are 99% identical and consist of 1140 bp, encoding 379 amino acids with no introns in the coding sequence [14, 64]. Sequence comparisons of these two homoeologous genes revealed four single nucleotide polymorphisms (SNPs) in the coding region among a low O/L and a high O/L line. Two of the four SNPs were synonymous; whereas the other two identified SNPs were characterized by an insertion generating a shift in the reading frame and a missense mutation, 442insA or G448A, respectively [27, 64]. Both of these mutations are assumed to generate a loss of one of the three histidine boxes, which are involved in the metal ion complex required for oxygen reduction [27, 64, 65].

The expression patterns of *ahFAD2A* and *ahFAD2B* have also been previously assessed. Digestion of RT-PCR products demonstrated that *ahFAD2A* is expressed in normal and high oleate seeds; however, *ahfad2B* had reduced transcript levels (or a lack of transcript) that correlated with the HO trait [14]. The presence of premature stop codons produced downstream from the 442insA in *ahfad2B* likely initiates cellular mechanisms for the degradation of aberrant mRNA via nonsense-mediated decay (NMD) to avoid accumulation of deviant proteins that might have adverse effects [66]. Furthermore, Jung et al. [63] established that the G448A missense mutation in *ahfad2A* and the reduction in transcript levels of *ahfad2B* were both necessary in the tetraploid peanut to produce the high oleate phenotype; whereas the normal oleate phenotype was generated by only a single gene expressing a functional enzyme. Another less commonly employed mutation to develop HO lines was derived from mutagenesis which resulted in a miniature inverted transposable element (MITE) inserted at 665 and 997 bp of the coding region in *ahFAD2B* [67]. The G448A missense mutation (in *ahFAD2A*) would still be required in lines developed with a MITE insertion in *ahFAD2B* to produce a HO peanut.

Since the mutations for a HO phenotype have been well characterized, traditional breeding in conjugation with marker assisted selection (MAS) has been an efficient method to select HO peanuts. Historically, traditional breeding was coupled with gas chromatography (GC) to determine which lines carried the HO trait. Yet, this requires that all lines must be grown out to produce seeds before any phenotyping data can be collected to assess the trait. There are currently multiple methods (using different platforms) to select desired genotypes in *ahFAD2*. Cleaved amplified polymorphic sequence (CAPS) markers [68, 69], real-time PCR genotyping [70, 71], and an allele specific PCR [AS-PCR] [72] have all been recently developed to select the HO trait in breeding programs. However, traditional breeding was utilized prior to the development of these marker systems to cultivate numerous HO peanut lines that were characterized by gas chromatography (GC). These marker systems, however, are extremely advantageous in breeding programs for several reasons. First of all, parents can be selected that contain the spontaneous natural G448A mutation in *ahFAD2A* ($ol_1ol_1Ol_2Ol_2$, normal oleic) to cross with a HO genotype ($ol_1ol_1ol_2ol_2$) which would produce a 3:1 segregation of normal to HO progeny in the F_2 generation instead of 15:1 segregation which significantly increases the number of HO progenies obtained. Also, peanuts are generally a self pollinated crop, so marker assisted selection (MAS) can be employed to help determine which progeny are true hybrids versus progeny that are a product of self fertilization by comparing the genotypes of the parents to the progeny. Lastly, the marker systems provide the rapid detection of the desired genotypes in each generation from DNA extracted from slices of non-germinated seed (cotyledon tissue) allowing the undesirable genotypes to be eliminated before planting. This saves valuable time and space required to grow out all the plants produced in segregating populations and wait until seeds are produced to evaluate the trait of interest.

High Oleic Acid Oilseed Crops and Production

In comparison with traditional (or normal) cultivars with low level of oleic acid, the level of oleic acid had been significantly enhanced. The results from five major oilseed crops are summarized in Table 2 and expressed in Figure 2. For example, the level of oleic acid in HO cultivars in cottonseed oil and soybean oil has increased three or four fold. However, at present, the newly developed techniques have not been freely integrated into breeding programs.

Additionally, the current growing areas for these major oilseed crops in the world have not been completely covered by HO cultivars. How to increase HO production is still challenging and in some cases there are specific needs to grow normal oleic cultivars. For example, in order to promote growing HO peanut cultivars, farmers and processors in Texas cooperate to only grow and process HO peanut cultivars. Hopefully, this kind of regulation can be put into effect worldwide to increase HO cultivars acreage for other oilseed crops. Certainly, different climates and areas need different types of cultivars for increasing grain yield. To meet this requirement, breeders need to develop more new HO cultivars and especially focus on developing HO cultivars with disease resistance. It is really a long process from cultivar development to consumer consumption. First, not all HO cultivars are available and cultivated by farmers. Second, not all the seeds contain HO acid are used to make products by processors. Third, not all products containing HO are available and consumed by consumers. There is a lot of work which need to be done for making the connection from breeders to farmers, from processor to consumers, and eventually offer HO acid for benefiting humankind.

Table 2. Fatty acid composition in high-oleic (HO) varieties from five major oilseed crops

Fatty acid/Species oil	Soybean [a]	Canola [b]	Sunflower [c]	Cottonseed [d]	Peanut [e]
Oleic acid (C18:1)%	85.1	84.0	79.0	78.2	78.9
Linoleic acid (C18:2)%	1.6	5.0	12.0	3.7	3.7
Linolenic acid (C18:3)%	3.2	3.0	-	0.1	-
Palmitic acid (C16:0)%	7.3	5.0	3.0	15.3	5.8
Stearic acid (C18:0)%	2.7	2.0	4.0	2.3	3.5
Others%	0.1	1.0	2.0	10.4	8.1

Sources: [a] [38]; [b] ftp://ftp.fao.org; [c] [48]; [d] [58]; and [e] [76].

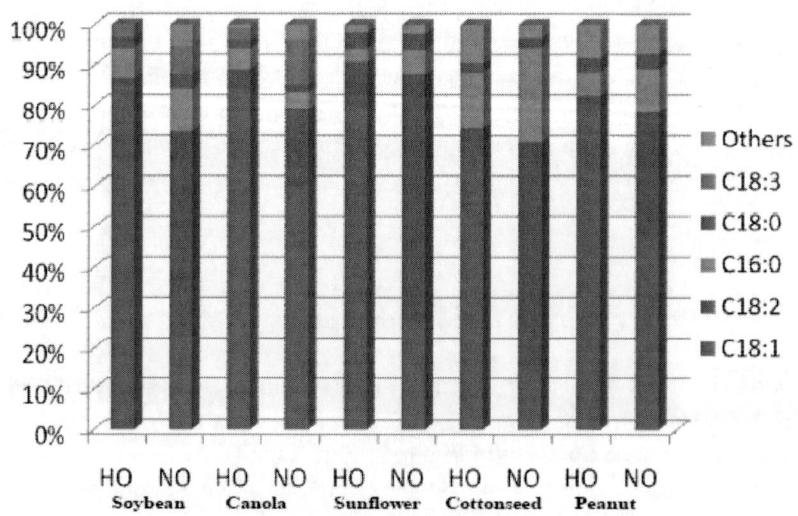

Figure 2. Comparison of fatty acid composition between high oleic (HO) and normal oleic (NO) cultivars among five major oilseed crops.

Conclusion

Researches demonstrated that consumption of oil containing high oleic acid or its derived products can benefit human health. However, the mechanism of how oleic acid promotes human health is not very clear and more basic research needs to be conducted for revealing its molecular function in human diet. Natural variation on the level of oleic acid exists in oilseed crops. Some of the natural variation in certain crops had been identified. Identifying genetic variation in oilseed crops allows the manipulation of the fatty acid profile through traditional breeding and marker-assisted selection (MAS). Molecular markers targeting mutant alleles for the HO trait allows the identification of the desired genotypes at early generational stage which can further reduce the time and effort to develop HO cultivars. However, molecular breeding will only provide limited changes to the overall fatty acid profile in many oilseed crops.

In the last two decades, oilseed crops have undergone significant changes in fatty acid composition through mutagenesis, RNAi, or genetic transformation. Mutation analysis has demonstrated that crop plants can withstand significant changes in lipid composition without serious effects to the survival of the plant. These methods, although time consuming and albeit fairly low efficiency rates in some crops, have proven to be successful in modifying fatty acid composition to obtain oils that are more stable than their unmodified counterparts. Furthermore, RNAi and genetic transformation are techniques that can modify multiple fatty acids simultaneously by targeting and modifying expression of more than a single gene linked to fatty acid accumulation producing novel fatty acid profiles. Progress and research in mutagenesis and transformation ultimately will allow oilseeds to be developed producing seeds with higher oil content, a nutritional balance of fatty acids, longer shelf life for stored food products and increasing or decreasing specific fatty acids for industrial purposes or health benefits.

References

[1] A. Kachroo, J. Shanklin, E. Whittle, L. Lapchyk, D. Hildebrand, and P. Kachroo, *Plant Mol.Biol.* 63, 257 (2007).
[2] D. T. Win, *Australian J. Tech.* 9, 75 (2005).
[3] F. D. Gunstone, J. L. Harwood, and A. J. Dijkstra, *The lipid handbook*, Taylor and Francis, Boca Raton, FL (2007).
[4] J. B. Ohlrogge, *Biochimica et Biophysica. Acta,* 1082, 1 (1991).
[5] A. Kachroo, S. C. Venugopal, L. Lapchyk, D. Falcone, D. Hildebrand, and P. Kachroo, *Proc. Natl. Acad. Sci. USA*, 101, 5152 (2004).
[6] Q. Wu, T. Liu, H. Liu, and G. Zheng, *African J. Biotech.* 8, 1782 (2009).
[7] E. E. Mosley, G. L. Powell, M. B. Riley, and T. C. Jenkins, *J. Lipid Res.* 43, 290 (2002).
[8] D. Zhang, L. L. Pirtle, S. J. Park, M. Nampaisansuk, P. Neogi, S. W. Wanjie, R. M. Pirtle, and K. K. Chapman, *Plant Phys. Biochem.* 47, 462 (2009).
[9] S. S. Funari, F. Barceló, and P. V. Escribá, *J. Lipid Res.* 44, 567 (2003).
[10] R. G. Upchurch, *Biotech. Lett.* 30, 967 (2008).

[11] N. Verdoni, M. Mench, C. Cassagne, and J. J. Bessoule, *Environ. Tox. Chem.* 20, 382 (2001).

[12] Q. M. Gao, S. Venugopal, D. Navarre, and A. Kachroo, *J. Amer. Oil Chem. Soc.* 84, 47 (2011).

[13] K. M. Moore, and D. A. Knauft, *J. Heredity*, 80, 252 (1989).

[14] S. Jung, D. Swift, E. Sengoku, M. Patel, F. Teule, G. Powell, K. Moore, and A. Abbott, *Mol. Gen. Genet.* 263, 796 (2000b).

[15] Y. López, O. D. Smith, S. A. Senseman, and W. L. Rooney, *Crop Sci.* 41, 51 (2001).

[16] O. Korver, and M. B. Kata, *Nutr. Reviews*, 64, 275 (2006).

[17] J. B. Ohlrogge, *Plant Physiol*, 104, 821 (1994).

[18] D. Mozaffarian, A. Aro, and W. C. Willett, *European J. Clin. Nutr.* 63, S5 (2009).

[19] M. I. Covas, V. Ruiz-Gutierrez, R. de la Torre, A. Kafatos, R. M. Lamuela-Raventos, J. Osada, R. W. Owen, and F. Visioli, *Nutr. Reviews*, 64, S20 (2006).

[20] P. Yaqoob, *Brazilian J. Medical Bio. Res.* 31, 453 (1998).

[21] S. Terés, G. Barceló-Coblijn, M. Benet, R. Álvarez, R. Bressani, J. E. Halver, and P. V. Escribá, *Proc. Natl. Acad. Sci. USA*, 105, 13811 (2008).

[22] L. Pelkman, V. K. Fishell, D. H. Maddox, T. A. Pearson, D. T. Mauger, and P. M. Kris-Etherton, *Am. J. Clin. Nutr.* 79, 204 (2004).

[23] S. Parthasarathy, J. C. Khoo, E. Miller, J. Barnett, J. L. Witztum, and D. Steinberg, *Proc. Natl. Acad. Sci. USA*. 87, (1990).

[24] J. O'Byrne, D. A. Knauft, and R. B. Shireman, *Lipids*, 32, 687 (1997).

[25] K. Vassiliou, A. Gonzalez, C. Garcia, J. H. Tadros, G. Chakraborty, and J. H. Toney, *Lipids Health Dis.* 8, 25 (2009).

[26] J. M. Martin-Moreno, W. C. Willett, L. Gorgojo, J. R. Banegas, F. Rodriguez-Artalejo, J. C. Fernandez-Rodriguez, P. Maisonneuve, and P. Boyle, *Intl. J. Cancer*, 58, 774 (1994).

[27] S. Yu, L. Pan, Q. Yang, P. Min, Z. Ren, and H. Zhang, *J. Genet. Genomics*, 35, 679 (2008).

[28] G. Knothe, J. van Gerpen, and J. Krahl, *The Biodiesel Handbook*, AOCS Press, Champaign, IL (2005).

[29] J. F. Battey, K. M. Schmid, and J. B. Ohlrogge, *Trends Biotech.* 7, 122, (1989).

[30] M. E. Tat, P. S. Wang, J. H. van Gerpen, and T. E. Clemente, *J. Am. Oil Chem. Soc.* 84, 865 (2007).

[31] C. Lu, J. A. Napier, T. E. Clemente, and E. B. Cahoon, *Curr. Opin. Biotech.* 22, 252 (2011).

[32] Q. Liu, S. Singh, and A. Green, *J. Am. Coll. Nutr.* 21, 205S (2002).

[33] P. C. Andersen, and D. W. Gorbet, *J. Ag. Food Chem.* 50, 1298 (2002).

[34] Z. Flagella, T. Rotunno, E. tarantino, R. Di Caterina, and A. De Caro, *European J. Agron.* 17, 221 (2002).

[35] T. G. Isleib, R. F. Wilson, and W. P. Novitzky, *Crop Sci.* 46, 1331, (2006a).

[36] A. Patil, S. P. Taware, M. D. Oak, S. A. Tamhankar, and V. S. Rao, *J. Am. Oil Chem. Soc.* 84, 1117 (2007).

[37] A. Martin, R. F. Wilson, and R. W. Rinne, *J. Am. Oil Chem. Soc.* 63, 346 (1986).

[38] T. Pham, J. D. Lee, J. G. Shannon, and K. D. Bilyeu, *Theor. Appl. Genet.* 123, 793 (2011).

[39] S. M. Rahman, Y. Takagi, K. Kubota, K. Miyamoto, and T. Kawakita, *Biosci. Biotech. Biochem.* 58, 1070 (1994).
[40] T. Buhr, S. Sato, F. Ebrahim, A. Xing, Y. Zhou, M. Mathiesen, B. Schweiger, A. J. Kinney, P. Staswick, and T. Clemente, *Plant J.* 30, 155 (2002).
[41] G. Graef, B. J. LaVallee, P. Tenopir, M. Tat, B. Schweiger, A. J. Kinney, J. H. van Gerpen, and T. E. Clemente, *Plant Biotech. J.* 7, 411, (2009).
[42] P. A. Stoutjesdijk, C. Hurlestone, S. P. Singh, and A. G. Green, *Biochem. Soc. Transactions.* 28, 938 (2000).
[43] X. Hu, M. Sullivan-Gilbert, M. Gupta, and S. A. Thompson, *Theor. Appl. Genet.* 113, 497 (2006).
[44] L. Auld, M. K. Heikkinen, D. A. Erickson, J. L. Sernyk, and J. E. Romero, *Crop Sci.* 32, 657 (1992).
[45] Tian, D. Sun, Y. Lian, H. Shu, H. Ling, X. Zang, B. Wang, and Z. Pei, *African J. Microbiol. Res.* 5, 817 (2011).
[46] J. H. Jung, H. Kim, Y. S. Go, S. B. Lee, C. Hur, H. U. Kim, M. and C. Suh, *Plant Cell Rep.* 30, 1881 (2011).
[47] W. Sabetta, V. Alba, A. Blanco, and C. Montemuro, *Plant Methods.* 7, 20 (2011).
[48] D. Škorić, S. Jocić, Z. Sakač, and N. Lečić, *Can. J. Physiol. Pharmacol.* 86, 215 (2008).
[49] M. Baldini, R. Givanardi, and G. P. Vanozzi, *In: Proc. XV Intl Sunflower Conf.* Toulouse A79, 84 (2000).
[50] K. Soldatov, *In: Proc. 7th Intl. Sunflower Conf.* Krasnodar, USSR 352-357 (1976).
[51] S. Lacombe, I. Souyris, and A. J. Bervillé, *Mol. Genet. Genomics* 281, 43 (2009).
[52] G. Nagaraj, *Peanut Sci.* 14, 7 (1993).
[53] U. Gecgel, M. Demirci, E. Esendal, and M. Tasan, *J. Amer. Oil Chem. Soc.* 84, 47 (2007).
[54] P. Golkar, A. Arzani, and A. M. Rezaei, *J. Amer. Oil Chem. Soc.* 88, 975 (2011).
[55] P. F. Knowles, *J. Amer. Oil Chem. Soc.* 49, 27 (1971).
[56] E. Ashrafi, and K. Razmjoo, *J. Amer. Oil Chem. Soc.* 87, 499 (2010).
[57] K. D. Chapman, S. Austin-Brown, S. A. Sparace, A. J. Kinney, K. G. Ripp, I. L. Pirtle, and R. M. Pirtle, *J. Amer. Oil Chem. Soc.* 78, 941 (2001).
[58] Q. Liu, S. P. Singh, and A. G. Green, *Plant Phys.* 129, 1732 (2002b).
[59] B. Uzun, C. Arslan, S. Furat, *J. Amer. Oil Chem. Soc.* 85, 1135 (2008).
[60] H. Yoshida, M. Tanaka, Y. Tomiyama, and Y. Mizushina, *J. Amer. Oil Chem. Soc.* 84, 165 (2007).
[61] B. Uzun, S. Ülger, M. I. Çağirgan, *Turk. J. Ag. For.* 26, 269 (2002).
[62] J. Norden, D. W. Gorbet, D. A. Knauft, and C. T. Young, *Peanut Sci.* 14, 7 (1987).
[63] S. Jung, G. Powell, K. Moore, and A. Abbott, *Mol. Gen. Genet.* 263, 806 (2000a).
[64] Y. López, H. L. Nadaf, O. D. Smith, J. P. Connell, A. S. Reddy, and A. K. Fritz, *Theor. Appl. Genetics.* 101, 1131 (2000).
[65] C. Bruner, S. Jung, A. G. Abbott, and G. L. Powell, *Crop Sci.* 41, 522 (2001).
[66] K. E. Baker, and R. Parker, *Curr. Opin. Cell Biol.* 16, 293 (2004).
[67] M. Patel, S. Jung, K. Moore, G. Powell, C. Ainsworth, and A. Abbott, *Theor. Appl. Genet.* 108, 1492 (2004).
[68] Y. Chu, L. Ramos, C. C. Holbrook, and P. Ozias-Akins, *Crop Sci.* 47, 2372 (2007).
[69] Y. Chu, C. C. Holbrook, and P. Ozias-Akins, *Crop Sci.* 49, 2029 (2009).

[70] N. A. Barkley, K. D. Chenault Chamberlin, M. L. Wang, and R. N. Pittman, *Mol. Breed*. 25, 541 (2010).

[71] N. A. Barkley, M. L. Wang, and R. N. Pittman, *Electronic J. Biotech*. DOI: 10.2225/vol14-issue1-fulltext-12 (2011).

[72] Z. Chen, M. L. Wang, N. A. Barkley, and R. N. Pittman, *Plant Mol. Biol. Rep*. 28, 542 (2010).

[73] Bils, R.F., and Howell, R.W. *Crop Sci.* 3, 304 (1963).

[74] J. A. Robertson, and V. E. Green, *J. Amer. Oil Chem. Soc*. 58, 698 (1981).

[75] P. J. van Niekerk, and A. E. C. Burger, *J. Amer. Oil Chem. Soc*. 62, 531 (1985).

[76] T. G. Isleib, H. E. Pattee, T. H. Sanders, K. W. Hendrix, and L. O. Dean, *J. Ag. Food Chem*. 54, 1759 (2006b).

In: Oleic Acid: Dietary Sources, Functions and Health Benefits ISBN: 978-1-62618-332-2
Editor: Luciano Paulino Silva

Chapter 3

VOLUME AND ADSORPTION PROPERTIES OF OLEIC ACID SOLUTIONS

Victor V. Korolev**, Anna G. Ramazanova, Olga V. Balmasova, Evgenii V. Ivanov and Valentina I. Yashkova

G.A. Krestov Institute of Solution Chemistry of the Russian Academy of Science, Ivanovo, Russia
Nanomagnetic Co., Ltd., Ivanovo, Russia

ABSTRACT

The present chapter covers the questions of experimental study of volume and adsorption properties of oleic acid on the surface of superfine magnetite. Oleic acid is the most common unsaturated fatty acid in nature; however considering of its properties is more interesting in comparison with its nearest homologous analogs (linoleic and linolenic acids) exhibiting intrinsic differences due to double bonds number. The limiting partial molar volumes of the solutes and excess molar volumes of their mixtures with nonpolar solvents were discussed. The above parameters were calculated from experimental data on density of dilute solutions of oleic, linoleic, and linolenic acids in tetrachloromethane, n-hexane, n-heptane, cyclohexane, and benzene at 298.15 K. High-precision vibration densimetry method was applied for density measurements; the measurements error did not exceed 5 10^{-6} g/cm^3. It was found that the processes of self-association of the mixtures components played the key role in dissolution effects of fatty acids in CCl_4. The increasing of C=C double bonds number in the fatty acid molecules make the acid desolvation less pronounced and was attended by compression (reduction of volume) of the structure packing of the solvatocomplex formed due to the strengthening interaction of the fatty acid with tetrachloromethane. It was shown that the solutions of fatty acids in *n*-alkanes became closer packed on going from oleic to linolenic acid and with decreasing in hydrocarbon chain length in the solvent molecules. This is caused, as a rule, by the enhancement of the *n*-alkane—acid intermolecular interaction. It was established that benzene influenced selectively on the volumetric

* Viktor V. Korolev, e-mail address: vvk@isc-ras.ru.

effects of dissolution of linoleic and linolenic polyunsaturated fatty acids. The adsorption properties of oleic acid and its analogs on the superfine magnetite surface were studied by equilibrium adsorption method. The sorption - desorption isotherms of oleic, linoleic or linolenic acids from their solutions in carbon tetrachloride, hexane, heptane or cyclohexane on the magnetite surface were obtained. The influence of the surfactants and the solvents on the adsorption properties was discussed. The Theory of Volume Filling of Micropores (TVFM) was applied to calculate the adsorption isotherms parameters, such as limiting adsorption, adsorption characteristic energy, and porous space volume, and their changes depending on surfactant and solvent nature were considered. The two-dimensional state diagrams of adsorption layers of the fatty acids on the solid surfaces were performed.

INTRODUCTION

During last two decades polyunsaturated fatty acids (PUFA) engross attention of researches all over the world. The most innovative investigations are devoted to study of the influence of a man feeding habits on development of immune and nervous systems, metabolism, and on visual functions formation. The role of polyunsaturated fatty acids in feeding of children and adults of different ages is of great interest. At that the most important data are ones demonstrating the high physiological activity of long-chain PUFAs of ω–3 type. It was established that these fatty acids participated in formation of brain cell membranes and neuroretina, in promotion of immune response, in anti-inflammatory reactions and other biological processes. Fatty acids (FAs) are carboxylic acids with long aliphatic chain and are termed FA because of the method of their isolation from fats. There are more than 800 natural fatty acids. However only nine main FAs, divided into 3 groups according to the unsaturation degree, will be marked out below:

1) saturated acids (e.g. palmitic ($C_{15}H_{31}COOH$) and stearic ($C_{17}H_{35}COOH$) acids);
2) monounsaturated acids (ω–9, e.g. oleic acid ($C_{17}H_{33}COOH$));
3) polyunsaturated acids (ω-3 and ω-6, e.g. linoleic ($C_{17}H_{31}COOH$), α-linolenic, γ–linolenic, arachidonic ($C_{19}H_{31}COOH$), eicosapentaenoic ($C_{20}H_{30}O_2$), and docosahexaenoic ($C_{22}H_{32}O_2$) acids) [1].

For human health, essential fatty acids play very important role as they can not be synthesized in human organism. It is well-known that many diseases are caused by deficiency of certain fatty acids and can be cured just by increasing of their content in dietary intake.

Oleic acid is undoubtedly the most widespread fatty acid in nature. Vegetable oils contain it as triglycerides: 70 – 85% of oleic acid is in olive oil, 75% – in almond oil, 74% – in palm kernel oil, and about 80% – in sunflower oil. There is oleic acid in butterfat of many animals, as well as in waxes and phosphatids. Some other fats and oils include from 5 to 40% of this fatty acid [1, 2].

The application of oleic acid and its salts is very extensive. They are used as components of detergents, drying oils, varnishes, emulsifiers, and flotators. Oleic acid esters are also applied as cellulose plasticizers and fragrants in food industry.

Linoleic acid is the main constituent of many vegetable oils. It is simple acid with two double bonds, which is found in many plant and animal tissues. This acid is necessary for growth, reproduction and healthy development but is not synthesized by animal tissues. Therefore linoleic acid must be entered a man organism mainly with food as complex lipids – triglycerides and phosphatids. There is considerable quantity (up to 40-60%) of linoleic and linolenic acids as triglycerides in many vegetable oils and animal fats, e.g. in soyabean, cottonseed, sunflowerseed, linseed (up to 30% of linolenic acid), hempseed, corn, and perilla (up to 55% of linolenic acid) oils, and in blubber. In animal tissues these acids are biosynthetic precursors of some other polyunsaturated fatty acids [3–5].

Minor amounts of linolenic acid can be synthesized from oleic and linoleic acids into plants and mammals, accordingly [2].

In [6] the composition of saturated and unsaturated FAs and steroids, entering into triglycerides of vegetable oils, was investigated by chromato- graphy-mass spectrometry method. The analysis of steroids composition allowed determining fats origin (vegetable or animal) by cholesterol content. As test samples the probes of purchased vegetable oils produced from soya, corn and sunflower were used. It was established that these oils included compounds of acid series from decanoic to tetracosanoic one. Unsaturated oleic and linoleic acids were main constituents. Small amount of linolenic acid was found out in soyabean oil only.

As it is generally known cardiovascular diseases are main causes of population mortality in many developed countries. Special attention of medics and politics is attracted to tendencies of morbidity and mortality because of cardiovascular diseases, some cancer types and other illnesses appeared in the last two decades of XX century. As main reasons of such situation the following factors are considered: economic and psychosocial conditions, as well as life-style including feeding, alcohol and smoking. Within the epidemiological investigations [7] it was discovered that feeding-style was the key factor which could explain the difference in risk levels of clinical atherosclerosis manifestations in various populations.

In [8] the relation between high-energy diet with large content of saturated fats and cholesterol and high risk of ischemic heart disease (IHD) was shown. In the work [7] the authors presented convincing data that the higher medium-population parameters for total blood cholesterol and consequently the IHD prevalence were registered in countries, where the feeding type with large amount of saturated fats formed due to geographic and natural living conditions and traditional handicrafts development. It was noticed that in the Mediterranean countries, where population consumed more vegetable food and fat component of diets was mainly composed of monounsaturated (MUFA) and polyunsaturated (PUFA) fatty acids and, in less degree, from saturated fats (SF), the mortality level because of IHD and ischemic stroke was considerably lower, moreover arterial hypertension and pancreatic diabetes of 2nd type were less distributed. Peculiarities of intake of vegetable food and fats in Europe and Japan are striking illustrations of feeding style significance. As it is known Japan and France take two last positions among other countries around the world concerning mortality level because of ischemic heart diseases. Economically these countries, like many others in Western Europe, USA and Canada, belong to advanced develop countries and environmental situation in Europe is almost the same for all countries. Main distinctions refer to the character of feeding traditions formed historically. The analysis carried out by authors [7] demonstrates that France and some other Mediterranean countries population, as against Northern and Eastern Europe, has diet peculiarities. There is less meat and more fish, a lot of verdure, orange and green vegetables in dietary intake in these countries. Moreover the fatty-acid composition of consumable fats is rather special. In Mediterranean countries the most popular are olive and nut oils, whereas in Japan it is soybean oil. According to the literature data [1] these oils are characterized by large content of α-linolenic fatty acid affecting positively the lipidic blood indexes. The feeding preferences, shifted toward larger fish and seafood and relatively less meat intake, act as antiatherogenic factor due to high content of ω-3 fatty acids.

In the work of Simonova G. I. [7] it was shown that optimal ration of mono- and polyunsaturated fatty acids, soya protein, dietary fibers, and antioxidant vitamins have cardioprotective and antiatherogenic effects. It was noticed that during Eskimos preventive treatment in Greenland the cardiovascular diseases connected with atherosclerosis, including acute cardiac infarction, were rather occasional [9, 10]. At rather large fats fraction in their daily food intake, as compared with feeding of other Scandinavians, ω-3 fatty acids prevailed in their diet [11-13].

According to up-to-date recommendations [14] prescription of hypolipide- mic drugs is integral part of modern prophylaxis and therapy of cardiovascular diseases with atherosclerotic genesis.

By some researchers [15] the complex investigations of physiologic peculiarities of fatty-acid blood composition and tocopherol concentration in blood of babies and teenagers living on the territory situated in different latitudes north were carried out. 242 residents of North Russia, including children and teenagers at the age 10-17 years, of northwestern area (Arkhangelsk region) and northeastern area (Anadyr, Chukotski autonomous region) were examined. By means of gas-liquid chromatography a number of fatty acids were determined in blood serum: 15 saturated (e.g. caproic, lauric, myristic, and stearic) acids, 9 mono-unsaturated (oleic) and 10 polyunsaturated (linoleic, linolenic and arachidonic) ones. Furthermore the children questioning on their food habits were conducted. The results obtained for practically healthy children living in different geographical latitudes point out that North-West population has lower level of polyunsaturated and higher concentrations of saturated or monounsaturated FAs. It is interesting that children and teenagers in North-Western region have down level of vitamin E, and it is one of the reasons of transformation of proteins, fats and carbohydrates metabolism mechanisms [15]. The study of actual children and teenagers food in Russia northern regions revealed the large difference in nutritional state and supply of fat-soluble vitamins and essential fatty acids. The imbalance of real nutrition of children and teenagers in the North was also discovered. Authors established that the average content of nutrients in the food did not fit with physiological needs of children under examination. Moreover the derangement of food state of children from North-Western regions is characterized by PUFA deficit, namely by vitamins lack, excess of simple carbohydrates and scarcity of dietary fibers (vegetables and fruits). The saturation with vitamin E of infantile organism depends not only on the vitamin content in food and intestine absorptive function, but also on the amount of PUFAs entering with food: the more such acids the higher the vitamin E needs. Consequently, high PUFAs (linolenic, eicosapentaenoic and docosahexaenoic) consumption with food by children in North-Western region increases vitamin E requirement. It was shown [15] that dietary intake of children in the North, especially in North-Western regions, did not provide sufficient amount not only of fat-soluble vitamins but also of proteins, essential ω-3 and ω-6 fatty acids. Primary power inputs of children living in this region are supplied not at the cost of fats but from bread, baked and corn goods. Thereby the reason of lower level of vitamin E and PUFAs in blood serum of children in North-Western region, as compared with children from other areas, is their unbalanced nutrition with predominance of food with high hydrocarbons content (bread, baked goods, dairy foods, and potato). In the diet of population of North-Eastern region food is rich in lipids and proteins (fish, seafood, venison).

Investigation of nutrition structure [15] revealed that children in North-Eastern region consumed a lot of seafood (northern fish in the first place) with high concentration of PUFAs and vitamin E. The data obtained for children strictly testify changes of serum fatty-acid composition with habitation latitude increasing. These changes were found to be tocopherol and PUFAs, mainly linolenic one, accumulation. At the same time children living in North-Western region had substantially increased concentrations of triglycerides and atherogenic lipid fractions attending by decrease of tocopherol content and growth of saturated and monounsaturated FAs concentrations. As it was stressed in [15] that the rise of oleic and pentadecanolic acids content can be undoubtedly considered as an important positive adaptive organism answering directed to triglycerides catabolism acceleration. The increase of monounsatu- rated FAs fraction is probably caused by peculiarities of spare triglycerides deposit or by synthesis of monounsaturated fatty acids in liver, as it can be connected with cholesterol etherification processes in blood channel.

Thus habitation in high latitudes is attended by activation of essential fatty acids synthesis in human organism. Taking into account the fact that the synthesis is regulable, it becomes obvious that there are some triggering mechanisms determined by difference in dietary.

UFAs influence not only on lipid metabolism but also on some human tissues, e.g. on lens. In [16] the results of adults examination are performed to discover some biochemical peculiarities of lenses of various mechanical strengths. At the time the lens examinations mainly aim at studying of their dimness mechanism, including very important processes of lipids peroxide oxidation. However, as authors [16] notice, now the lens mechanical strength is of great importance and understanding of the processes resulting in the lens strengthening is one of the key problems. The investigations conducted revealed the extent and direction of influence of unsaturated fatty acids on the lens mechanical strength. The direct relation between UFAs concentration and lens strength was discovered. So it was shown that concentrations of oleic, linoleic and linolenic acids in the cataract lens were rather low and fairly correlated (correlation coefficient 0.5) with strength of lens nucleus. It was concluded that the main reason of mechanical lens strengthening was accumulation of secondary products of peroxide lipids oxidation, namely of malonic dialdehyde, which formed dense insoluble polymer compound. All these findings corroborate once again the conclusion made earlier on necessity of good balanced diet not only for children but for adults also.

Plant and animal drugs, including considerable amounts of oleic, linoleic and linolenic fatty acids (vegetable oils, phospholipidic concentrates), are able to stabilize cell membrane. Besides they can normalize metabolic liver function and possess antioxidant activity. On the basis of experimental data some authors showed the availability of drugs containing essential fatty acids for toxic liver injury treatment [17]. So the black nut oil was used for toxic hepatitis cure.

Fatty acids play important role in many technological processes, e.g for magnetic liquids (MLs) synthesis. In MLs, the particles of magnetic phase are modified by fatty acids to prevent these particles aggregation. Hence, the questions of experimental study of volume properties of oleic acid solutions and the acid adsorption characteristics on the superfine magnetite surface are important issue of current days. Oleic acid is the most extended UFA in nature. However to get more comprehensive information on this fatty acid properties the comparison with its nearest homologous/analogs (linoleic and linolenic acids) is required.The latter fatty acids differ from oleic acid by double bonds number. In the present chapter the results of experimental investigation of volume properties of oleic, linoleic and linolenic acids solutions in carbon tetrachloride, hexane, heptane, cyclohexane, and benzene, as well as of adsorption properties of stated solutions on the superfine magnetite surface are performed.

VOLUME PROPERTIES OF OLEIC ACID AND ITS POLYUNSATURATED HOMOLOGUES IN SOLUTIONS

Unsaturated fatty acid (UFA) molecules are known to be indispensable and active participants of metabolism and biochemical synthesis including pharmacological processes [18, 19]. The UFAs forming homological series with a general structural formula $C_{18:n}$ (n is number of C=C bonds in the cis-9, cis-6, and cis-3 positions relative to the end methyl group), namely, oleic ($n_{C=C}$ = 1), linoleic ($n_{C=C}$ = 2), and linolenic ($n_{C=C}$ = 3) acids, are the major representants of this class of organic compounds. All of these UFAs possess strong antioxidant properties, the latter two (polyunsaturated acids) belonging to so-called essential fatty acids form vitamin F group [19]. However, purposeful study of the properties of biologically active media involving these fatty acids is restrained by their extremely low solubility in water.

At present time the information on structural and thermodynamic characteristics of liquid binary "organic solvent – fatty acid" systems takes on special significance. Therefore any additional experimental data on the properties of the fatty acids solutions in nonaqueous solvents, including the fatty acids solvation characteristics, will undoubtedly contribute to more clear understanding of the processes taking place in the solutions and on the phase interfaces. Today in literature sources, the results on volume properties of the above fatty acid solutions in carbon tetrachloride [20], n-hexane and n-heptane [21], in cyclohexane and benzene [22], as well as data on solution enthalpy of oleic and linoleic acids in carbon tetrachloride and in *n*-hexane at 298.15 K [23] were performed.

The state-of-the-art on densimetric technique, as it was shown above, allows obtaining very important information on volume and "packing" parameters connected directly with structural reorganization within the solvent under the influence of solute molecules [24, 25]. Moreover, analyzing the volumetric effects of solvation using semi-empirical model approximations, one can quantitatively estimate changes in the parameters of intermolecular interaction in binary liquid systems containing isomeric analogs or compounds from the same homological series as solutes are.

The experimental density values of UFAs solutions along with description of procedures of the solvents preparation, densimetric measurements, and calculation of volume characteristics one can find in references [20-22]. However in the present work only the data on volume properties of UFAs stated above both in pure and dissolved state in nonpolar organic media will be considered. The data on the volumetric properties of pure UFAs under study are listed in Table 1.

Table 1. Densities (ρ_2), molar volumes (V_2), and Van der Waals volumes ($V_{w,2}$) of the $C_{18:n}$ fatty acids under study at 298.15 K

UFA	Molecule structure	ρ_2, g/cm³	V_2,[(a)] cm³/mol	$V_{w,2}$, cm³/mol
Oleic	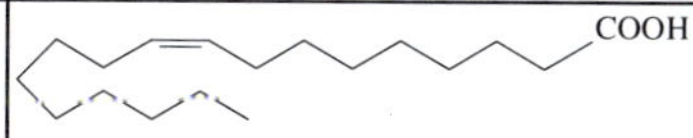	0.895593 $\pm 2\cdot 10^{-5}$	315.40	180.3
Linoleic	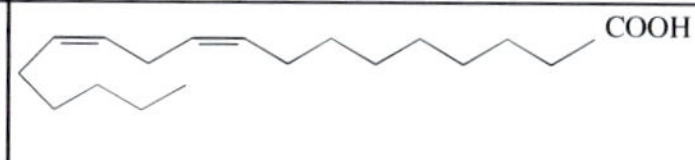	0.902702 $\pm 2\cdot 10^{-5}$	310.68	177.3
Linolenic	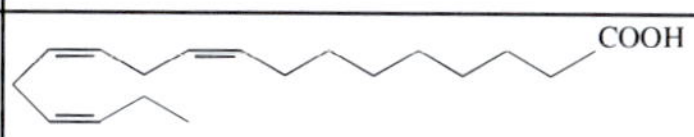	0.917132 $\pm 2\cdot 10^{-5}$	303.59	174.3

(a) $V_{w,2} = v_{w,2} N_A$, where NA is the Avogadro number and $v_{w,2}$ is the Van der Waals volume of the UFA molecule calculated according to taking into consideration the overlapping of the valence co-spheres of neighboring atoms.

Table 2. Volumetric properties (cm³ mol⁻¹) of infinite diluted solutions of unsaturated fatty acids in carbon tetrachloride ($\overline{V}_2^{\infty}, \overline{V}_2^{E,\infty}$) at 298.15 K

UFA	$V_{w,2}/V_2$ (a)	$\overline{V}_2^{\infty}$ (b)	$V_{w,2}/\overline{V}_2^{\infty}$	$\overline{V}_2^{E,\infty}$ (c)	$\overline{V}_2^{\infty} - V_1$ (d)
Oleic acid	0.571	321.33	0.561	5.93	224.28
Linoleic acid	0.571	314.32	0.564	3.64	217.27
Linolenic acid	0.574	304.18	0.573	0.59	207.13

(a) The volume properties of pure UFAs are listed in Table 1.

(b) The partial molar volume of acids at infinite dilution.

(c) $\overline{V}_2^{E,\infty} = \overline{V}_2^{\infty} - V_2$ is the excess limiting partial molar volume (or "volume effect of dissolution").

(d) The molar volume of carbon tetrachloride: V_1 = 97.049 (cm³ mol⁻¹).

Since the studied nonpolar solvents and the fatty acids solutions in these solvents differ significantly by structural and thermodynamic properties it is reasonable to discuss every system under investigation separately.

Solutions of Oleic and Polyunsaturated Fatty Acids in Carbon Tetrachloride

The volume properties of the UFAs solutions in carbon tetrachloride are listed in Table 2.

The data in Table 2 show that molecular packing of UFAs in the individual state ($V_{w,2}/V_2$ ratio) becomes, in general, more compact with $n_{C=C}$ increasing. This tendency also takes place for infinite diluted solutions of the acids in carbon tetrachloride. The only difference is that $V_{w,2}/\overline{V}_2^{\infty} < V_{w,2}/V_2$ or $\overline{V}_2^{E,\infty} > 0$ (Table 2), and that passing from oleic to linolenic acid the structure of UFA–CCl_4 solvate complexes compacts more significantly (for $n_{C=C} = 3$, $\overline{V}_2^{E,\infty}$ becomes even close to zero, i.e., $\overline{V}_2^{\infty} \approx V_2$). Also it should be noted that in given (diluted) region of the mixture compositions, the transfer of UFA molecules from their own liquid medium to the more close packed carbon tetrachloride medium leads to loosening of the latter, which increases with $c_{sm,2}$ and within the acids series: $C_{18:3} \rightarrow C_{18:2} \rightarrow C_{18:1}$. This can be seen from the character of the concentration dependences of excess molar volumes $V_{1,2}^{E}$ (Figure 1).

Taking into account the structural features of organic compounds under consideration (Table 1), one can state that changes in packing during formation of the extremely diluted acid solutions in CCl_4 (Table 2) are directly related to differences in the ability of UFA molecules to build into the structural matrix of the solvent. The relationship between the energy characteristics of CCl_4–CCl_4 (1-1) and CCl_4–UFA (1-2) interactions is one of the most important problems for infinitely dilute solutions as contacts between solute molecules (2-2) are excluded there.

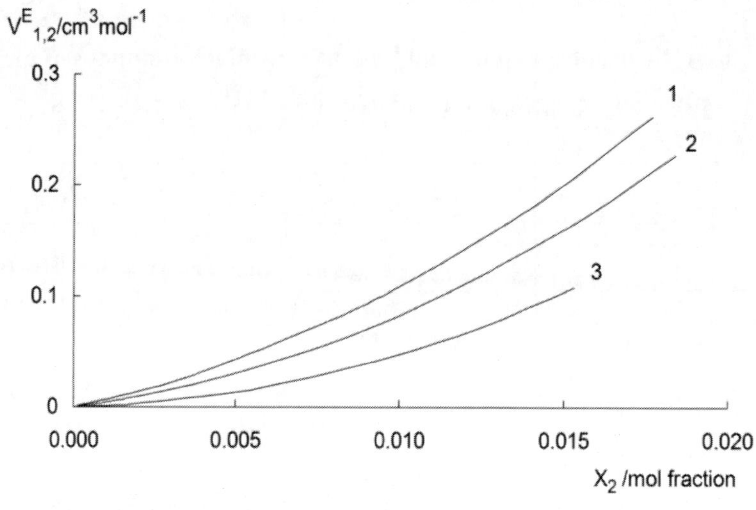

Figure 1. Concentration dependence of excess molar volumes of dilute (relative to the acid) solutions of oleic (1), linoleic (2), and linolenic (3) acids in carbon tetrachloride at 298.15 K (x_2 is an acid molar fraction in solution) (cited with permission).

To solve partly this problem one can use the approach suggested in [29], which relates the relative molecular affinity parameter $\Delta G^{*,\circ}$ to the volumetric properties of the binary system as following:

$$\Delta G^{*,\circ}/(kT\rho_1^*) = V_1 \Delta G^{*,\circ}/(RT) = G_{11}^{\circ} - G_{12}^{\circ} = \overline{V}_2^{\infty} - V_1, \quad (1)$$

Here k is the Boltzmann constant; $\rho_1^* = N_A/V_1$ is the number density; G_{11}° and G_{12}° are group integrals from the limiting expressions of Kirkwood–Buff theory for molar volume

[30, 31]. The values $\Delta G^{*,\circ} = RT(\bar{V}_2^{\infty} - V_1)/V_1$ for CCl_4 – UFA (kJ mol^{-1}) are equal to 5.73 ($C_{18:1}$), 5.55 ($C_{18:2}$) and 5.29 ($C_{18:3}$).

It follows from Equation (1) and from the data in Table 2 that $\bar{V}_2^{\infty} - V_1 >> 0$ and $\Delta G^{*,\circ} >> 0$, i.e. $G_{11}^{\circ} >> G_{12}^{\circ}$ (in absolute value). Thus we can conclude that the affinity of solvent molecules to one another prevails over the affinity of the solvent to UFA molecules. In other words, the volumetric changes found in solutions are caused by very low solubility of fatty acids $C_{18:n}$ in carbon tetrachloride. As individual UFAs are inclined to self-association (mainly forming dimers) [32], one can assume that this process dominates in the region of real acid concentrations close to infinite dilution.

At the same time, the tendency toward reduction of the positive effect $\Delta G^{*,\circ}$ (due to the partial replacement of CCl_4–CCl_4 bonds by weaker UFA–CCl_4 bonds) with $n_{C=C}$ increasing in acid molecules points to more strong solvation of these molecules (dimers decomposition) and formation of closer packed structures with surrounding solvent molecules. This is confirmed by increasing of exothermic effect [23] of UFA dissolution in carbon tetrachloride when passing from oleic to linolenic acid. Thus at T = 298.15 K and $x_2 \approx 1.5 \cdot 10^{-5}$ the solution enthalpy $\Delta_{sol}H_2^c(CCl_4)$ is 9.26 kJ mol^{-1} for oleic acid and 6.86 kJ mol^{-1} for linolenic one. Taking into account the above changes in the volumetric effects of UFA solution, one can assume that $\Delta_{sol}H_2^c$ is within the interval from 4 to 4.5 kJ mol^{-1} for the linolenic acid - CCl_4 system.

More in-depth information on structural transformations in the liquid systems can be obtained from model representations allowing one to divide the total (integrated) volumetric characteristics of UFA solvation into separate components. One of these approaches is statistical-mechanical calculation of contribution from extremely dilute solution of the fatty acids in tetrachloromethane to the volumetric properties. Such calculations base on approximation of component molecules by hard sphere models, specifically within the scale particle theory (SPT).

Although this theory [33-35] includes some formal assumptions, it is based on the experimentally measured thermodynamic (including volumetric) properties of solvents such as isothermal compressibility $\beta_{T,1} = (\partial V_1/\partial p)_T$. The expression for the latter is an exact solution of Percus–Yevik state equation [33]. Moreover, hypothetical replacement of a "real" polyatomic molecule by an equivalent in volume hard sphere with a diameter σ seems to be quite justified in the case of CCl_4. For "long-chain" UFAs in question (Table 1) it allows quantitative evaluation of relative changes in packing during solvation of substances belonging to the same homological series.

According to SPT, the expression for the partial molar volume of a solute in infinitely dilute solution is following:

$$\bar{V}_2^{\infty} = \bar{V}_{cav}^{\infty} + \bar{V}_{int}^{\infty} + \beta_{T,1}RT \quad (2)$$

where $\overline{V}_{\text{cav}}^{\infty}$ is a term due to the formation of solvate cavity in the liquid solvent and $\overline{V}_{\text{int}}^{\infty}$ is a term due to the intermolecular 1-2 interactions; $\beta_{T,1}RT$ is a contribution taking into account a change of the system standard state during the "gas → liquid" transition.

The value of $\overline{V}_{\text{cav}}^{\infty}$ was obtained using the following equation [34]:

$$\overline{V}_{\text{cav}}^{\infty} = \beta_{T,1}RT\left[\frac{y}{1-y}+\frac{3yz(1+z)}{(1-y)^2}+\frac{9y^2z^2}{(1-y)^3}\right]+\frac{\pi\sigma_2^3 N_A}{6} \quad (3)$$

Here y is the packing parameter calculated from the relation $\beta_{T,1} = V_1(1-y)^4/[RT(1+2y)^2]$ $z = \sigma_2/\sigma_1$, where σ_1 and σ_2 are hard-sphere diameters of the solvent (CCl_4) and solute (UFA) molecules, respectively. The expression $y = \pi\sigma_1^3 N_A/(6V_1)$ [33] is used to calculate σ_1, and the formula $\sigma_2 = [6V_{w,2}/(\pi N_A)]^{1/3}$ [18] is applied for calculating σ_2.

Table 3 presents the parameters of Equations (2) and (3) and the volumetric contributions to $\overline{V}_2^{\infty}$ calculated using these equations for CCl_4–UFA liquid systems.

Analyzing the data from Table 3 the following conclusions can be drawn. Firstly, the $\overline{V}_{\text{cav}}^{\infty}$ value decreases monotonously (by ~4.2 cm^3/mol) as the number of C=C double bonds in the UFA molecule increases. Secondly, the positive volumetric contribution $\overline{V}_{\text{int}}^{\infty}$ becomes by half less passing from oleic to linolenic acid. The difference between the values of $\overline{V}_{\text{int}}^{\infty}$ in the stated series of UFA also decreases twice when linolenic acid is replaced by linoleic one. These circumstances confirm the above assumption that, on the one hand, the solvation ability of carbon tetrachloride relative to UFA molecules is rather weak and, on the other hand, the tendency toward stronger interactions between CCl_4 and UFA molecules becomes especially pronounced while comparing the different polyunsaturated fatty acids.

Table 3. Parameters of Equation (2) and volumetric characteristics of the components of extremely diluted solutions of unsaturated fatty acids in carbon tetrachloride at 298.15 K [20] necessary for these parameters calculation

UFA	y	$\sigma_1 \times 10^8$, cm	$\sigma_2 \times 10^8$, cm	z	$\beta_{T,1}RT$, (a) kJ/mol	$\overline{V}_{\text{cav}}^{\infty}$, cm^3/mol	$\overline{V}_{\text{int}}^{\infty}$, cm^3/mol
Oleic A	0.4398	5.135	8.300	1.616	2.70	300.5	18.1
Linoleic A	0.4398	5.135	8.254	1.607	2.70	296.3	15.3
Linolenic A	0.4298	5.135	8.207	1.598	2.70	292.1	9.4

(a) $\beta_{T,1} \approx 10.91 \cdot 10^{-10}$ Pa^{-1}.

It should be kept in mind that planar UFA molecules differ slightly in the size of the hydrocarbon chain and in polarizability values α_0. For oleic and linolenic acids α_0 values were estimated by Lorenz-Lorenz equation using data on refractive index and V_2 from [37], and amounted to 33.2×10^{-24} cm^3/mol and 34.7×10^{-24} cm^3/mol at 293.15 K, respectively. Subject to the stated above Van der Waals (mainly dispersion) intermolecular interactions are not likely to be the reason for the revealed packing (structural) transformations in liquid CCl_4 induced by solution of fatty acids $C_{18:n}$ therein. From the viewpoint of the UFAs structure it is most substantiated to assume, as it was made in [18, 19], the dominant role of C=C double bonds in the structure of acid–solvent molecular solvatocomplexes. It is not excluded that π-σ-interactions with the solute environment lead to compression of the stretched spatially UFA molecule (or change its structural conformation), which increases with $n_{C=C}$. This is also favored by the electron acceptor ability of CCl_4, which is higher than the electron acceptor abilities of some other nonpolar aprotic solvents [38].

Solutions of Oleic and Polyunsaturated Fatty Acids in Hexane and Heptane

As it can be seen from the data in Tables 1 and 3, the structure packing of UFAs in the individual state are rather loose unconsolidated molecular aggregates ($V_2 >> V_{w,2}$) becoming more compact with $n_{C=C}$ increasing. The results presented in the Table 4 demonstrate that the difference between intrinsic molar volume of linolenic acid and the limiting partial molar volumes of this acid in *n*-hexane (*n*-C_6H_{14}) or *n*-heptane (*n*-C_7H_{16}) does not exceed 5.7%, and for heptane solution of oleic acid the $\overline{V}_2^{E,\omega}$ value even is close to zero (i.e., $\overline{V}_2^{\infty} \approx V_2$). This indicates that the volumes of free ("excluded") space near the acid molecule, which is surrounded by intrinsic molecules and nonpolar solvent molecules, are comparable.

Table 4. Volume properties (cm^3/mol) of infinite dilute solutions of the unsaturated fatty acids in n-alkanes at 298.15 K

n-Alkane	UFA	$\overline{V}_2^{\infty}$	$-\overline{V}_2^{E,\omega}$
n-Hexane	Oleic A	303.22	12.18
	Linoleic A	295.37	15.31
	Linolenic A	286.23	17.36
n-Heptane	Oleic A	314.97	0.43
	Linoleic A	307.73	2.95
	Linolenic A	300.01	3.58

At the same time, the regular increase in the negative $\overline{V}_2^{E,\infty}$ values in the series of oleic, linoleic, and linolenic acids indicates an increasing compression of the molecular packing of the acid—solvent solvate complexes formed. A similar tendency is observed when replacing heptane by hexane in the solvate environment of dissolved UFA molecules (see Table 4).

The decrease of $\overline{V}_2^{\infty}$ with shortening of the hydrocarbon chain length in the n-alkane molecules is related, most likely, to the loosening and compressibility increasing of the solvent structural packing around the UFA molecules on going from heptane to hexane. This weakens the interaction between the solvent molecules. According to the earlier published data [39] the polarizability $\alpha_{0,1}$ and force constant ϵ_{11}/k of the Lennard-Jones pair potential (12—6) for hexane and heptane molecules are $11.83 \cdot 10^{-24}$ cm^3 and 517 K, and $13.70 \cdot 10^{-24}$ cm^3 and 573 K, respectively.

The evaporation enthalpies of n-alkanes in question are 31.55 and 36.55 kJ/mole [37], and the isothermic compressibilities $\beta_{T,1}$ are $16.27 \cdot 10^{-10}$ and $14.09 \cdot 10^{-10}$ Pa^{-1} [36, 40] for hexane and heptane, respectively. It is most likely that the observed packing changes upon the acids dissolving are directly related to differences in both the solvation properties of the solvents and the ability of "bulky" (long-chain) solute molecules to incorporate into the initial structural matrix of the solvent.

As it was mentioned above, the UFAs under study dissolved in hexane endothermically. This effect became less pronounced with increasing in the $n_{C=C}$ in the solute molecules. For instance, on going from oleic to linoleic acid, the solvation enthalpy $\Delta_{sol}H_2^c$ ($x_2 \approx 1.5 \cdot 10^{-5}$) the acids in n-C_6H_{14} decreases from 8.8 to 6.2 kJ/mol [23]. This trend being attended by the negative $\left|\overline{V}_2^{E,\infty}\right|$ values increasing (see Table 4) enables to assume that upon $n_{C=C}$ growth in the acid molecules as well as when heptane being replaced by hexane the acids solvation enhances. As there are no data on heat effects of UFAs dissolution in heptane one can not confirm or refute the assumption stated above. However for this purpose the model conceptions within the TVFM expressed by equations (2) and (3) (see section 1.1) can be used.

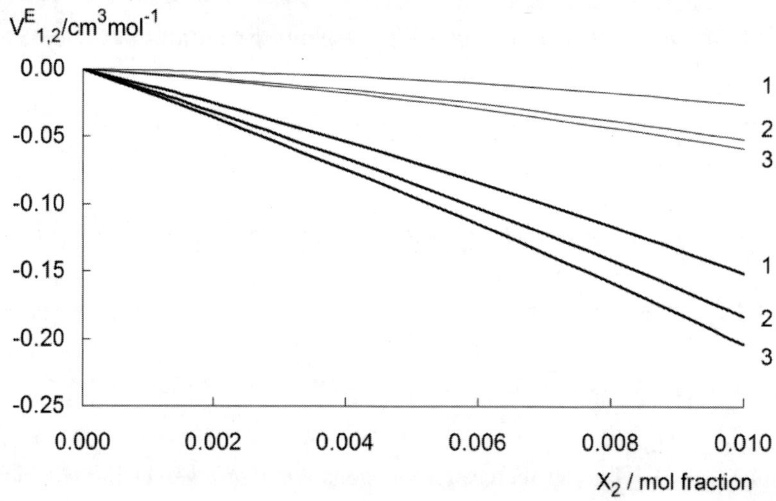

Figure 2. Concentration dependences of the excess molar volumes of n-alkane—unsaturated fatty acid mixtures for oleic (1), linoleic (2), and linolenic (3) acids at 298.15 K: solid lines – UFAs solutions in hexane; dash lines - UFAs solutions in heptane.

As it can be seen from the data in Table 5, the positive volume contribution to $\overline{V}_2^{\infty}$ coming from the structural cavity formation in the solvent is compensated by the negative contribution $\overline{V}_{\text{int}}^{\infty}$. The last value in its absolute expression increases sharply on going from heptane to hexane and within the acid series: oleic → linoleic → linolenic. The difference in volumes $|\overline{V}_{\text{int}}^{\infty}(C_6H_{14}) - \overline{V}_{\text{int}}^{\infty}(C_7H_{16})|$ on going from oleic to linolenic acid enhances from 15.6 to 17.5 cm^3/mol.

Table 5. Parameters of Equation (2) and volume characteristics (required for the calculation of these parameters) of the components of infinite dilute solutions of unsaturated fatty acids in n-alkanes at 298.15 K

Solvent	UFA	y	$\sigma_1 \times 10^8$, cm	$\sigma_2 \times 10^8$, cm	$\overline{V}_{\text{cav}}^{\infty}$, cm^3/mol	$-\overline{V}_{\text{int}}^{\infty}$, cm^3/mol
Hexane	Oleic A	0.4295	5.637	8.300	319.6	20.4
	Linoleic A	0.4295	5.637	8.254	315.2	23.9
	Linolenic A	0.4295	5.637	8.207	310.8	28.6
Heptane	Oleic A	0.4572	5.979	8.300	316.3	4.81
	Linoleic A	0.4572	5.979	8.254	311.9	7.70
	Linolenic A	0.4572	5.979	8.207	307.6	11.1

[a] The procedure of the parameters presented in the Table calculation is given in the section 1.1.

Taking into account the magnitudes of "parameters of intermolecular interaction" in the solvents in question, the solvating power of heptane medium should be higher. However, the incorporation of spatially extended UFA molecules into the more closely packed and less compressible structural heptane matrix is fraught with additional difficulties of the configurational character, that is indicated, in particular, by an increase in $\overline{V}_2^{\infty}$ in this solvent (see Table 4). The increasing role of a steric incompatibility during the formation of UFA – n-alkane solvate complex is a probable reason for the weakening of intermolecular interactions 1–2 on going from hexane to heptane UFA solutions.

Additional information on the packing effects in the solutions under study can be obtained from analysis of their excess molar volumes $V_{1,2}^{E}$ shown in Figure 1.

As one can see in Figure 2 the increase in the UFA content in the solution results in the formation of more closely packed structures. Such tendency becomes more pronounced in the series of oleic acid → linoleic acid → linolenic acid and on going from heptane to hexane solutions of these acids. According to [27, 41] under the mixing of liquids with different chemical structure and geometric shape of their molecules, configurational effects play determining role in the processes of structural packing formation and solvation of the solutes. Taking this fact into consideration the evident correlation of behavior of the data performed in Figure 2, and in Tables 4 and 5 can be regarded as a confirmation of our conclusions.

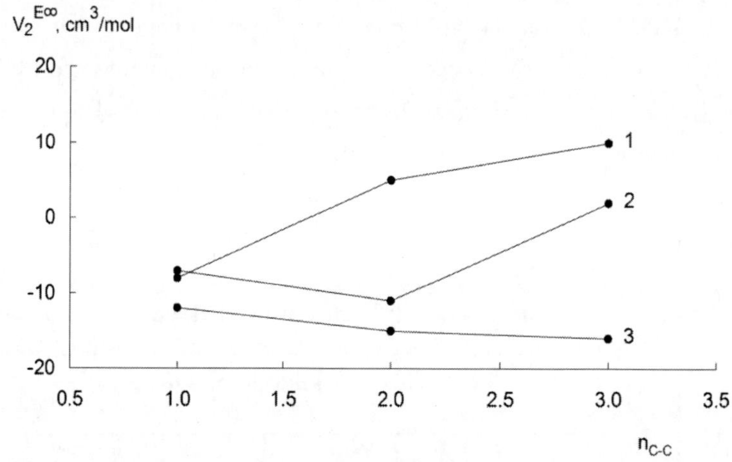

Figure 3. Dependence of the excess limiting partial molar volume of $C_{18:n}$ UFAs on the number of C=C bonds in their molecules for the solutions of these acids in (1) cyclohexane, (2) benzene, and (3) n-hexane at 298.15 K.

Solutions of Oleic and Polyunsaturated Fatty Acids in Cyclohexane and Benzene

The analysis of the data in Tables 1 and 6 shows that the values of $\overline{V}_2^{\infty}$ for the studied UFAs in c-C_6H_{12} and C_6H_6 differ by 5% at the most (for linolenic acid) from the acids molar volumes. This suggests that the volumes of the free (or so-called excluded) space near an UFA molecule surrounded by its own molecules or by molecules of a nonpolar solvent are similar.

At the same time the absence of any regularity in $\overline{V}_2^{\infty}$ changes for the systems in question (Table 6) allows to suppose that the configurational effects associated with the features of the solvents molecular structure play a key role in the formation of the structural packing of UFA–solvent solvatocomplexes. This is confirmed by the data on the $\overline{V}_2^{\infty}$ variation for UFAs of C18:n series dissolved in n-hexane, a linear hydrocarbon (Table 4), and by the values of the excess limiting partial molar volume for these UFAs in cyclohexane and benzene (Figure 3). The above discussion and the results presented in the Figure 3 led us to the following conclusions.

First, for oleic acid only, the $\overline{V}_2^{E,\infty}$ value is negative; its modulus decreases within the row n-C_6H_{14} → c-C_6H_{12} → C_6H_6, i.e. on passing to "cyclic" solvents and with decreasing of saturation degree in the solvents molecules. Second, the replacement of the linear saturated hydrocarbon (n-C_6H_{14}) as a solvent by a cyclic hydrocarbon (c-C_6H_{12}) produces a differentiating effect on the direction of $\overline{V}_2^{E,\infty}$ changes with $n_{C=C}$ increasing in the dissolved UFA molecule. For example, while for the n-C_6H_{14} solvent the molecular packing of the acid–solvent solvatocomplex progressively compresses on passing from oleic to linolenic acid, for the c-C_6H_{12} solvent it loosens for the same series of acids up to a state corresponding to $\overline{V}_2^{E,\infty} > 0$ at $n_{C=C} > 2$. Third, benzene as a solvating medium exhibits a kind of selectivity with respect to polyunsaturated fatty acids. The Figure 3 shows that the oleic acid → linoleic acid replacement in a benzene-based solution is accompanied by the formation of more tightly packed ($\delta\overline{V}_2^{E,\infty} \approx -5$ cm^3/mol) UFA–C_6H_6 structures; on the other hand, the changeover to linolenic acid results in the positive-to-negative inversion of the volumetric effect of the acid dissolution.

The last circumstance suggests that the molecular packing of benzene, comprised of planar-conjugated cycles, is the most suited structure for incorporating in it linoleic acid long-chain molecules (Table 1), exhibiting at that additional steric mismatches during the formation of linolenic acid–C_6H_6 solvatocomplexes of similar type. It is interesting fact observed in Figure 3 that the "abnormal" difference between the values of $\overline{V}_2^{E,\infty}$ for linoleic acid in c-C_6H_{12} and C_6H_6 (≈17 cm^3/mol) is close modulo to the same value $\delta\overline{V}_2^{E,\infty}$ (*n*-C_6H_{14}→C_6H_6) for linolenic acid. This can be explained by difference in the solvation ability of hydrocarbons with respect to polyunsaturated fatty acids. However to clarify this question the missing data on enthalpy effects of UFAs dissolution (solvation) of $C_{18:n}$ series in the solvents under investigation are required.

It is known that the heat of vaporization and molecular packing density increase while the compressibility and molecular polarizability decrease in the solvents row: *n*-C_6H_{14} → *c*-C_6H_{12} → C_6H_6 [36]. Note that the hydrocarbon chain lengths [18] and polarizabilities (vide supra section 1.1) do not differ significantly for planar UFAs molecules of $C_{18:n}$ series. Based on these data and the dependences displayed in the Figure 3, we concluded that the Van der Waals intermolecular interactions could not be considered as the main reason of cyclohexane and benzene structural (packing) rearrangements under the action of polyunsaturated fatty acids dissolved in these solvents.

It is more reasonable, in our opinion, that the structure of the UFA–solvent complexes is largely determined by the contribution from C=C bonds. It was demonstrated (section 1.2.) that the increase in $n_{C=C}$ favors the formation of denser solvatocomplexes during the incorporation of fatty acid molecules into the branched-chain structure of n-hexane. On passing to cyclohexane, the liquid with a lower compressibility, the opposite effect is observed (Figure 3). It is possible that, when placed in benzene, the elongated UFA molecules compress (or change their structural conformations) under the action of π-π and π-σ interactions with the solvation environment; the magnitude of these changes increases with $n_{C=C}$. The presence of such directional contacts between dipole and π-electrons of the benzene ring is supported by the data reported in [34, 42]. This is promoted by the donor–acceptor ability of C_6H_6 being higher than that of the other C_6 hydrocarbons [36].

At the same time on $\overline{V}_2^{E,\infty}$ dependence versus $n_{C=C}$ (Figure 3) the inversion, observed on passing from linoleic acid to linolenic one in benzene solution, implies that these interactions occur against a background of significant configurational rearrangements in the UFA–nonpolar solvent system.

In the Table 7 the results of modeling of interaction parameters for UFA – cyclohexane/benzene systems within the TVFM using equations (2) and (3) are listed.

A comparison of the data presented in the Figure 3, in Tables 5 and 7 allows drawing two main conclusions at least. First, only for solutions of the UFAs in n-C_6H_{14} the negative volumetric effects of the fatty acids dissolution ($\overline{V}_2^{E,\infty} < 0$) are accompanied by similar (in sign and direction) changes in $\overline{V}_{int}^{\infty}$ with number of C=C bonds increasing, which reflects the tendency to the strengthening of the solutes solvation. For solutions of the fatty acids in cyclohexane and benzene, except for the c-C_6H_{12} – oleic acid system, the key effects are the self-association of the components or the UFA molecules desolvation of ($\overline{V}_{int}^{\infty} > 0$ and $\overline{V}_2^{E,\infty} > 0$), which are generally most pronounced in C_6H_6 solutions. Second, it is clear that the above effect of benzene selectivity with respect to linoleic acid, and partly to linolenic one, is not associated with a significant changes in the character of the acid– C_6H_6 interaction; it rather stems from changes in the conformation of long-chain UFA molecules as they are incorporated into the close packed solvent medium.

Table 6. The limiting partial molar volumes (cm³/mol) of the UFAs in cyclohexane and benzene at 298.15 K

Acid	*c*-Hexane	Benzene
Oleic A	306.24	308.37
Linoleic A	315.66	298.66
Linolenic A	313.34	305.26

Note: The values of $\overline{V}_2^{\infty}$ were calculated within an accuracy ±0.1 cm³/mol.

Table 7. Parameters of Equation (2) and the volumetric characteristics of the components of infinite dilute solutions of the UFAs in cyclohexane and benzene required for calculating these parameters

System	y	$\sigma_1 \times 10^8$	$\sigma_2 \times 10^8$	$\beta_{T,1}RT$	$\overline{V}_{cav}^{\infty}$	$\overline{V}_{int}^{\infty}$
		cm		cm³/mol		
c-C_6H_{12}–Oleic A	0.4392	5.331	8.300	3.05	306.3	-3.1
c-C_6H_{12}–Linoleic A	0.4392	5.331	8.254	3.05	302.2	10.4
c-C_6H_{12}–Linolenic A	0.4392	5.331	8.207	3.05	297.7	12.6
C_6H_6– Oleic A	0.4473	5.025	8.300	2.33	294.6	11.4
C_6H_6– Linoleic A	0.4473	5.025	8.254	2.33	290.5	5.8
C_6H_6– Linolenic A	0.4473	5.025	8.207	2.33	286.2	16.7

At the end of this section it should be noted that the volume thermodynamic properties of solutions bring some information on macroscopic or supramolecular changes in binary systems, but they do not explain unambiguously the nature of structural reorganization on molecular level in the solvent under the solute influence. However the results obtained show that volume thermodynamic properties make it possible to perform both qualitative and quantitative analysis of packing effects induced by dissolution (solvation) of nonelectrolytes with close structure, in our case of unsaturated fatty acids of $C_{18:n}$ series in alkanes. Moreover, as it will be demonstrate in the section 3, the investigation of volume thermodynamic properties is of great importance for adsorption processes modeling, specifically within TVFM.

ADSORPTION PROPERTIES OF OLEIC ACID ON THE MAGNETITE SURFACE

Some conclusions can be drawn as the result of the work performed. Firstly, during UFAs dissolution in carbon tetrachloride medium self-association processes of the mixture components play dominating role for dissolution effects. At that the larger the double bonds (C=C) number in the acid molecule the less pronounced the acid desolvation. Secondly, the endothermic effect of the acids dissolution decreases in the acids row: oleic → linoleic → linolenic, and is accompanied by compression (volume decreasing) of structural packing of UFA – CCl_4 solvatocomplex formed. The last circumstance is of great importance for adsorption processes modeling, specifically within the TVFM.

To study the influence of surfactant and solvent nature on adsorption processes, the adsorption of oleic, linoleic and linolenic acids from solutions with carbon tetrachloride, hexane, cyclohexane or heptane on the magnetite surface was carried out by the technique described in [23, 43] Equilibrium-adsorption method was applied to obtain the adsorption isotherms, from which the surfactant adsorption parameters were calculated. Over the region of equilibrium concentrations these parameters obey the Theory of Volume Filling of Micropores.

Earlier [44, 45] it was established that there was a certain competition between solute and solvent molecules for adsorbent surface active centers. Moreover it was revealed that the number of double bonds in fatty acids molecules played a key role, as well as desolvation of adsorbent surface and adsorbate molecules in nonpolar solvent media, in the process of fatty acids adsorption [46].

For more comprehensive description of experimental adsorption isotherms the plane state diagrams of surface layers were drawn to obtain the complementary information on the state of adsorbed substances in surface layers over the whole equilibrium concentrations range.

It was shown [47] that the aggregate state of the adsorbate surface layer changed in the course of adsorption. These variations can be characterized by the set of state parameters and equations. The last may be applied to description of adsorbed substances state in adsorption volumes of solid surfaces.

According to [48] the adsorbate state in adsorption volumes of solids is similar to the substance state in surface films. At that the shape of plan state diagrams is the same as of ones for surface films. State diagrams enable to select equations for description of adsorbed substances state in surface layers and to prove the possibility of their structure transformation caused by two-dimensional phase transitions. As a state equation for surface layers one can use two-dimensional analogs of well-known state equations for bulk phases.

Two-dimensional state diagrams of surface layers were obtained on the base of experimental adsorption isotherms. The plotting of such diagrams is founded on the Gibbs integral (1). Equation (1) allows calculating two-dimensional surface pressure (π) depending on molar adsorbate area (S_m) which in turn equal to the reciprocal adsorption value (*a*) [49]:

$$\pi = RT \int_0^{C_i} A_i d \ln C_i \qquad (1),$$

where A_i is an adsorption value at a solute concentration C_i.

Molar area (S_m) was determined by equation:

$$S_m = 1/a \tag{2}$$

The values of two-dimensional surface pressures π required for plotting of the diagrams were founded by numerical integration of experimental adsorption isotherms using the procedure of smoothing spline - interpolation within experimental errors bar by Simpson`s and trapezoids methods. The estimated inaccuracy of calculated π values did not exceed 3÷5 % [48].

Thus using experimental and literature data [23, 49] it can be concluded that the experimental isotherms of oleic, linoleic, and linolenic acids adsorption from organic solutions on magnetite surface (Figure 4a, b) are of L-type according to Giles classification. Such isotherms characterize adsorption with parallel orientation of adsorbate molecules relative to the surface. Moreover the processes of reorientation of adsorbate molecules within surface layers are possible under the solute concentration increasing and desorption of solvent molecules from adsorbent surface. It is typical for this isotherm type that there is almost level plateau in the region of low equilibrium concentrations, that may corresponds to a certain structure of surface layers. Further concentration rising on the adsorption isotherms conforms to the condensation - type piece [48, 49].

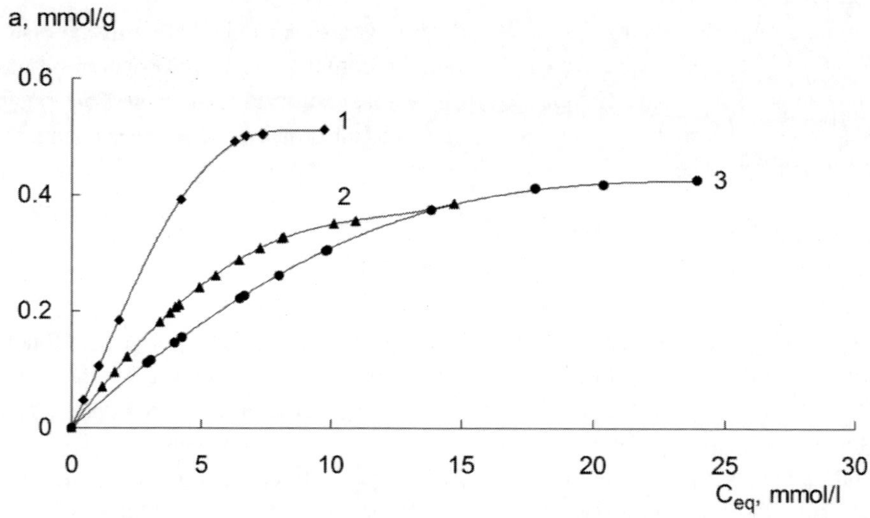

A

Figure 4. Continued on next page.

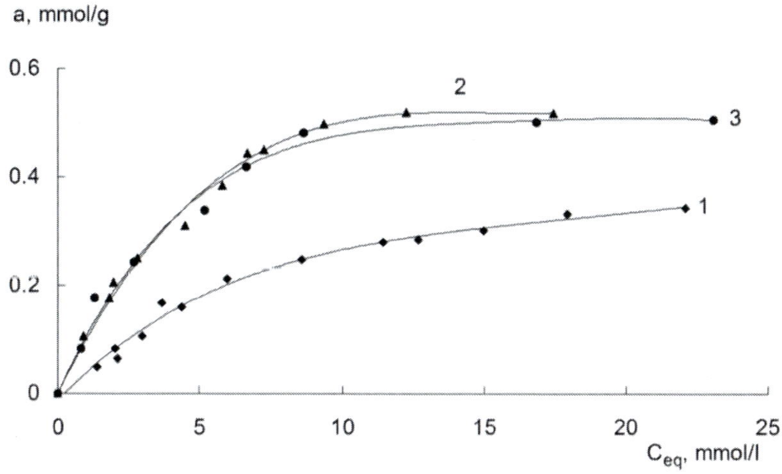

b

Figure 4. Adsorption isotherms of fatty acids from solutions with cyclohexane (a) and heptane (b) on the magnetite surface: ♦ (1) – oleic acid, ▲ (2) – linoleic acid, ● (3) – linolenic acid.

The main concentration feature of a surface layer is an adsorption value. Therefore the experimental adsorption isotherm is an objective two-dimensional characteristic of the state of real adsorbate – adsorbent system. It implies that any process within the adsorbent surface layer will change the form of two-dimensional state diagrams of this layer in the course of adsorption.

As an example of real system state diagrams in Figure 5(a, b) the state diagrams ($\pi = f(S_m)$) of oleic, linoleic and linolenic acids in cyclohexane and heptane solutions on the magnetite surface are presented.

As one can see in Figure 5 two-dimensional state diagrams of the fatty acids in surface layers are sister ones and similar to the state of real surface films [48]. Calculations demonstrates that even at low equilibrium concentrations the adsorption values exceeds monolayer filling of the surface, and the adsorption isotherms do not fit any known adsorption models within the wide concentration range.

Analyzing the experimental data obtained certain conclusions can be drawn on structural peculiarities of magnetite surface layers under the fatty acids adsorption from organic solvents. It was established that in all state diagrams there were two regions with different mechanisms of surface layers formation.

Over the region of high molar areas (S_m), corresponding to low concentrations on adsorption isotherms for magnetite 0÷5 mol/l (Figure 4), the diagram type is similar to the one for real gas surface films [31]. The results obtained argue that within this concentration range the adsorption processes proceed by mechanism of volume filling of ferrites porous space with adsorption solutions, and adsorption equilibria can be described by TVFM equation for microporous adsorbents [23, 50].

At higher solutes concentrations, corresponding to adsorption values higher than 5 mol/l (Figure 4), the strength of adsorbate-adsorbent and adsorbate-adsorbate interactions changes in adsorption solutions. To a certain extent this fact can be confirmed by appearance of inflections on the isotherms (Figure 4) and of curve pieces (Figure 5, marked) like to the ones corresponding to two-dimensional phase transitions in surface films [48]. The existence of concentration range, which corresponds to two-dimensional phase transition, on the state diagrams can be considered as an evidence of adsorption layers rearrangement with solutes concentration increasing.

With further rising of adsorption value, e.g. for linolenic acid adsorption from heptane at concentrations 10-25 mmol/l (Figure 4b), on the state diagrams one can observe the region which conforms to the one emerging when condenced liquid surface films exist. It is obvious that over this concentration range surface layers form by the mechanism of condensation interactions and the concentration dependence of adsorption values is not fitted by any classical adsorption model [48, 50].

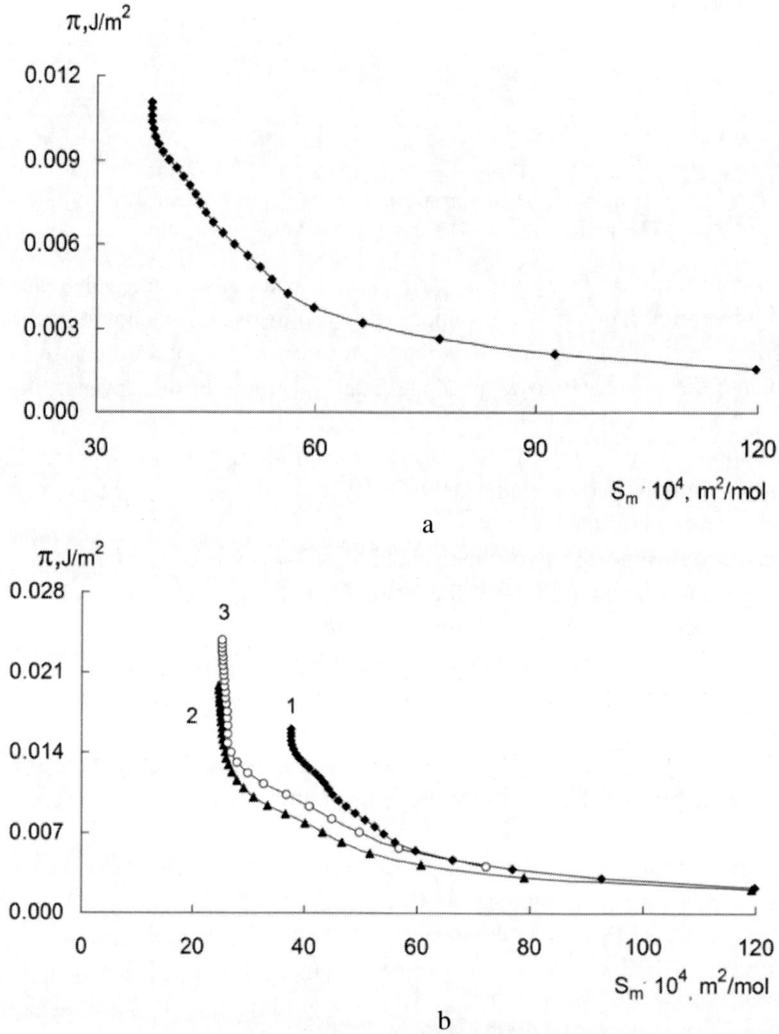

Figure 5. Two-dimensional state diagrams of fatty acids in organic solutions: ♦ – oleic acid from cyclohexane solutions (a); ♦ – oleic acid (1), ▲ (2) – linoleic acid, and ○ (3) – linolenic acid from heptane solutions (b) on the magnetite surface.

Moreover one of the probable reasons of the structural changes in the surface layers with adsorption increasing is the adsorbent deformation [50], which brings to formation of two-dimensional phase transition region on the state diagram and to inflection of adsorption isotherms.

Thus the existence of different regions on the state diagrams confirmes changes of surface layers formation mechanism with rising of the adsorbates concentration in solutions.

CORRELATION OF ADSORPTION ISOTHERMS PARAMETERS OF OLEIC AND POLYUNSATURATED FATTY ACIDS WITH VOLUMETRIC EFFECTS OF THEIR DISSOLVING IN ORGANIC SOLVENTS

In the previous section the fact of competition between solute and solvent molecules for the adsorbent surface-active centers was revealed during investigation of adsorption processes of oleic, linoleic and linolenic acids from solutions in *n*-hexane and carbon tetrachloride on the superfine magnetite surface.

It was established that the adsorption of aforementioned unsaturated fatty acids was mainly affected by the number of double bonds in their molecules, as well as by the desolvation of adsorbent surface and adsorbate molecules. With allowance for the latter circumstance the following fact is of special interest that in two different solvating medium, n-C_6H_{14} and CCl_4, the sequence of positions of the acid adsorption isotherms changes. For example, upon the adsorption from carbon tetrachloride solutions the values of limiting adsorption a_m, characteristic energy of adsorption E_a, and volume of porous space V_p of magnetite increase in oleic – linoleic - linolenic acids series, but all these parameters decrease in the same acid series under adsorption from hexane solutions.

Evidently, the effect of the apolar solvent nature on the processes on the ferroparticles surface is directly related with the specific features of intermolecular interaction between the adsorbate and solvation environment. At macroscopic level these features are exhibited in thermodynamic characteristics of dissolution (solvation) of UFAs. The most important among them, as it was shown in Section 1, are the volumetric or packing effects that depict structural changes in the solvent due to desolvation of acid molecules and formation of UFA-n-C_6H_{14} (CCl_4) salvatocomplexes of different sizes (volumes) and compositions.

Therefore the elucidation of correlations in changes of adsorption and partial volumetric characteristics of studied UFAs in n-hexane and tetrachloride should promote the understanding of processes that are responsible for the stability and efficiency of different magnetite-containing colloidal systems. The most interesting correlations of this kind were obtained in [51] during investigation of oleic and some other (polyunsaturated) fatty acids solutions in carbon tetrachloride and hexane. Her we would like to consider some results of that work.

It follows from the data presented in Tables 2 (Section 1.1) and 4 (Section 1.2) that in the individual liquid state 1 mole of molecules of each acid in question occupies smaller volume (positive values of $\overline{V}_2^{E,\infty}$) than in CCl_4. On the contrary, the dissolution of UFAs in n-hexane is accompanied by the formation of solvatocomplexes with more close-packed structure (negative values). Obviously, in UFAs solutions in hexane solvation of the solute molecules dominates, while, in carbon tetrachloride, their self-association, or desolvation, is predominant. At that in both non-polar solvents the tendency to increasing of UFA molecules solvation (or to decreasing of their desolvation) is found for acid series of oleic – linoleic – linolenic one.

This conclusion is confirmed by the previous results of calorimetric studies of oleic and linolenic acids solutions in carbon tetrachloride and n-hexane [52]. In particular it was shown that with $n_{C=C}$ increasing in the acid molecules the endothermic effect of the acid dissolution in carbon tetrachloride and n-hexane decreases by 2.40 and 1.83 kJ/mol, respectively. The transfer enthalpies $\Delta_{tr}H_2^c$ (n-C_6H_{14} → CCl_4) for comparable UFAs are negative (–1.18 kJ/mol at $n_{C=C}$ = 1 and –0.61 kJ/mole at $n_{C=C}$ = 2) that points out the higher salvation ability of hexane. However, as the number of double bonds in $C_{18:n}$ acid increases, this difference becomes less pronounced.

Note that the positive sign of $\Delta_{sol}H_2^c$ value, in essence, corresponds only to decrease in the solute solvation during its transfer from the own medium to the solvent and does not allow to judge unambiguously about the nature of interaction with solvation environment [24]. From this standpoint, taking into account the data of Table 2 and 4, volumetric characteristics of solutions seem to be more informative.

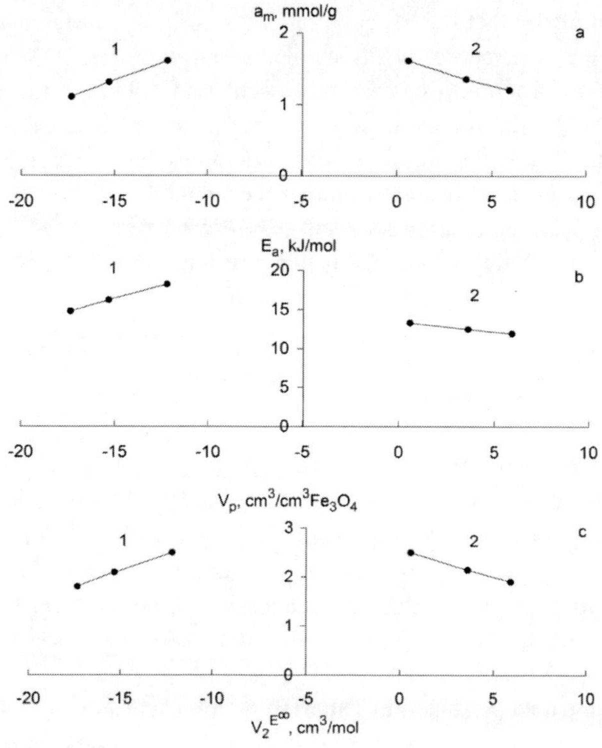

Figure 6. Dependences of various parameters of UFAs of $C_{18:n}$ series adsorption on the magnetite vs. volume effect of UFAs dissolution in n-hexane (1) and carbon tetrachloride (2): limiting adsorption (a), characteristic adsorption energy (b), and values of adsorbent porous space (c).

Indeed, carbon tetrachloride and n-hexane, as mutually dissolving media, are characterized by the close values of evaporation enthalpy, polarizability of molecules, and parameters of Lennard-Jones pair interaction potentials [36]. At the same time, the values of their compressibility differ considerably: β_T values for n-C_6H_{14} and CCl_4 are equal to $16.27 \cdot 10^{-10}$ and $10.91 \cdot 10^{-10}$ 1/Pa, respectively [36]. With regard to this, we can assume that more close-packed and less compressible medium of carbon tetrachloride is less suitable for the incorporation of "long-chain" UFA molecules into the solvent structure (see Table 1) despite of its higher electron–acceptor ability [24, 36]. In other words, volumetric effects because of steric hindrances, arising upon dissolution, are determining factors in the formation of $C_{18:n}$ fatty acids solutions in n-C_6H_{14} and CCl_4.

As it is seen from the Figure 6, the regularities revealed for the changes in the volumetric properties of oleic, linoleic, and linolenic acid solutions in n-hexane and carbon tetrachloride are mainly responsible for the selectivity of UFA molecules adsorption on the surface of superfine magnetite from these solvents.

When analyzing dependences shown in the Figure 6, one can come to the following conclusions. First, when UFA is dissolved in apolar medium with positive volumetric effect, the efficiency of the filling of micropores on the adsorbent surface increases with the number of double bounds in the adsorbate molecule or with enhancement of its solvation. That is the CCl_4 medium is more preferable for linolenic acid ($n_{C=C}$ = 3) adsorption on the magnetite. When the volume effect of UFA dissolution is negative (in n-hexane), the adsorptivity, on the contrary, appears to be much higher for the adsorbate with minimal, within studied acid series, $n_{C=C}$ or the volumetric dissolution (solvation) effect.

Second, by the character of changes in a_m and V_p values (see the Figure 6a and 6c), the linoleic acid ($n_{C=C}$ = 2) adsorptivity is formally independent of solvent nature. However, current difference in the characteristic adsorption energy of this and other acids (Figure 6b) suggest that an increase in the partial volume (loosening of the structural packing of solvatocomplex) of UFA upon its transfer from n-C_6H_{14} to CCl_4 is the main limiting factor in the process of micropores filling on the magnetite surface with acid molecules.

The latter conclusion is supported by the results of the approximation of dependences presented in the Figure 6 by linear equation $Y = A + B\overline{V}_2^{E,\infty}$, where Y = a_m, E_a, or V_p. Coefficient A corresponds to the dissolution of hypothetical UFA with zero $\overline{V}_2^{E,\infty}$ value, i.e. to the state when $V_2 = \overline{V}_2^{\infty}$. It is evident that such solvation properties should characterize the acid for which $n_{C=C} < 1$ for n-C_6H_{14} and $n_{C=C} > 3$ for CCl_4. The a_m (mol/g), E_a (kJ/mol), and V_p (cm^3/cm^3 Fe_3O_4) values calculated for the case $\overline{V}_2^{E,\infty} = 0$ are equal to 2.6 (±0.1), 13.4 (±0.1), and 1.6 (±0.2) for the CCl_4–UFA system and 4.1 (±0.9), 26 (±5), and 2.8 (±0.1) for the n-C_6H_{14}–UFA system. Comparing these parameters, we can state that the adsorptivity of fatty acids and the affinity of magnetite to UFAs of $C_{18:n}$ series are higher in hexane (especially for oleic acid) despite the lower solvation ability of carbon tetrachloride.

These data prove convincingly the necessity for considering the role of volumetric effects of UFA dissolution when studying their adsorption on ferroparticle surface from solutions in apolar solvents and, as a consequence, when selecting dispersion medium for the synthesis of stable and highly efficient magnetite-containing colloidal systems.

Conclusion

Investigations of the volumetric thermodynamic properties of solutions that result in information on macroscopic or supramolecular changes in binary systems certainly fail to give an unequivocal explanation of the nature of structural transformations occurring in a solvent under the influence of a solute at the molecular level. At the same time, our results show that by investigating these characteristics it is possible to carry out not only qualitative, but also quantitative analysis of packing effects induced by dissolution (solvation) of structurally related nonelectrolytes (in this case, of unsaturated fatty acids of $C_{18:n}$ series) in apolar solvents. The following major conclusions may be drawn from this work. First, self-association of mixture components is dominant among the dissolution effects of the UFAs in apolar solvents. Desolvation of acids becomes less appreciable with C=C double bonds number increasing. Second, decreasing of endothermic effect of the acids dissolution within the series of oleic→linoleic→linolenic acid is accompanied by compression (reduction of volume) of the structural packing of the resulting UFA - solvents complex. The latter circumstance is of critical importance for modeling adsorption processes (in particular, within the framework of the theory of volume filling of micropores), and also for analyzing the efficiency of UFA selection (as surfactants on the surface of ferroparticles) to create stable high-tech magnetic liquids in future.

On the base of experimental adsorption isotherms of the solutes two-dimensional state diagrams of surface layers were plotted. At certain concentrations in the systems the changes in surface layer composition were observed. So over the region of high molar areas (S_m) (low solute concentrations) the diagrams type is similar to the one for real gas surface films. At higher solute concentrations in adsorption solutions the strength of adsorbate-adsorbent and adsorbate-adsorbate interactions changes. Here curve pieces like to the ones corresponding to two-dimensional phase transitions in surface films can be seen. With further adsorption value rising on the state diagrams one can observe the region which conforms to the liquid condenced films on the surface.

References

[1] Vas'kovskii V. E. Lipids. *Sorosovskii obrazovatelnii journal,* 3, 32 (1997).

[2] Biologicheskii enciklopedicheskii slovar'; Editor, Gilyarov M. S.; M.: *Sovetskaya Enciklopediya,* 365, (1986).

[3] Dmitrichenko M.; Pilipenko T. *Tovarovedenie and ekspertiza pischevih jirov, moloka and molochnih produktov;* Sankt- Peterburg.: Piter, 352. ISBN 5-94723-680-X1 (2003).

[4] Himicheskii enciklopedicheskii slovar'; Editor, Knunyanc I. L.; M.: *Sovetskaya Enciklopediya,* 407 (1983).

[5] *Bolshaya sovetskaya encyclopaedia:* v 30 tomah; M.: Bolshaya sovetskaya enciklopediya, (1969-1978).

[6] Hasanov V. V.; Pijhova G. L. et al. *Himiya rastitel'nogo sir'ya,* 3, 27 (2006).

[7] Simonova G. I. *Bulleten' SO RAMN.,* 120, 2, 80 (2006).

[8] Keys A. *Coronary heart disease in seven countries,* 1, 41, 211 (1970).

[9] Dueberg *J. Am. Clin. Nutr.,* 28, 958 (1975).

[10] Hansen J. C. *Arctic Medical Research*, 53, 1, 4 (1994).
[11] Medkova I. L. Vse o vegetarianstve; *M.: Mir*, 287 (1992).
[12] Perova N. V. *Kardiovaskulyarnaya terapiya i profilaktika*, 4, 101 (2005).
[13] Holman R. T. *Fed. Proc.* 23, 1062 (1964).
[14] Galyavitch A. S.; Salahova L. P. *Ateroskleroz and dislipidemii*, 1, 18 (2011).
[15] Bichkaeva F. A.; Tret'yakova T. V. et al. *Ekologiya cheloveka*, 3, 44 (2010).
[16] Chuprov A. D.; Kudryavceva Yu. V. et al. *Vestnik OGU*, 78, 779 (2007).
[17] Esaulenko E. E.; Ladutiko A. A.; Bykov I. M. *Allergologiya and immunologiya*, 11, 3, 260 (2010).
[18] Lehninger A. L. *Biochemistry. The Molecular Basis of All Structures and Functions;* Worth: New York (1972).
[19] Shilina M. N.; Kon I. Ya. *Vopr. Detsk. Dietol.* 2, 25 (2004).
[20] Ramazanova A. G.; Korolev V. V. et al. *J. Struct. Chem.* 47, 6, 1091 (2006).
[21] Ramazanova A. G.; Yashkova V. I. et al. *Russ. Chem. Bull., Internation. Edit.* 55, 666 (2006).
[22] Ramazanova A. G.; Korolev V. V. et al. *Russ. J. Phys. Chem.* A. 81, 4, 560 (2007).
[23] Korolev V. V.; Ramazanova A. G. et al. *Colloid J.* 66, 5, 784 (2004).
[24] Abrosimov V. K.; Korolev V. V. In *Eksperimental'nye metody khimii rastvorov:* Densimetriya, viskozimetriya i drugie metody; Editor, Kutepov A. M.; Nauka: Moscow (1997).
[25] Abrosimov V. K.; Ivanov E. V. *Teoreticheskie i Eksperimental'nye Metody Khimii Rastvorov;* Editor, Tsivadze A. Yu.; Prospect: Moscow (2011).
[26] Kuz'min V. S.; Katser S. V. *Bull. Russ. Acad. Sci., Div. Chem. Sci.* 41, 720 (1992).
[27] Belousov V. P.; Panov M. Y. *Thermodynamic Properties of Aqueous Solutions of Organic Substances;* Editor, Boca Raton; CRC Press: FL/Tokyo (1994).
[28] Ivanov E. V. *Russ. J. Phys. Chem.* 78, 1225 (2004).
[29] Ben-Naim A. *Solvation Thermodynamics;* Plenum Press.: New York (1987).
[30] Kirkwood J. G.; Buff, F. P. *J. Chem. Phys.* 19, 774 (1951).
[31] Ivanov E. V.; Abrosimov V. K. *Biologicheski Aktivnye Vechshestva v Rastvorakh: Struktura, Termodinamika i Reaktsionnaya Sposobnost';* Editor, Kutepov A.M.; Nauka: Moscow (2001).
[32] Grabovskii Yu. P.; Ivanova O. I. et al. *Magn. Gidrodin.* 3, 27 (1987).
[33] Stolypin V. F.; Mishustin A. I. Zh. *Fiz. Khim.* 61, 3226 (1987).
[34] Sakurai M. *Bull. Chem. Soc. Jpn.*, 63, 1695 (1990).
[35] Klofutar C.; *Nemec T. J. Sol. Chem.* 25, 1151 (1996).
[36] Marcus Y. *Ion Solvation*; Chichester: Wiley (1985).
[37] Rabinovich V. A.; Khavin Z. Ya. *Kratkii Spravochnik po Khimii;* Khimiya: Leningrad (1991).
[38] Mamontov G. *Chemistry of Nonaqueous Solutions: Current Progress*; Editor, Popov A. I.; VCH: New York (1994).
[39] Brükl N.; Kim J. I. *J. Phys. Chem.* 126, 133 (1981).
[40] Malhotra R.; Woolf L. A. J. *Chem. Thermodyn.* 23, 49 (1991).
[41] Tager A. A.; Adamova L. V. *Russ. Chem.* 49, 618 (1980).
[42] Karlstrom G.; Linse P. et al. *J. Amer. Chem. Soc.* 105, 377 (1983).
[43] Korolev V. V.; Balmasova O. V. *Russ. J. Phys. Chem.* 83, 6, 1018 (2009).
[44] Balmasova O. V.; Korolev V. V. et al. *J. Phys. Chem.* 84, 1, 76 (2010).

[45] Barbov A. V.; Ulitin M. V. *Russ. J. Phys. Chem.* 71, 12, 2237 (1997).
[46] Fridrihsberg D. A. Kurs kolloidnoi ximii; *Ximiya: L.,* 368 (1984).
[47] Adamson A. Fizicheskaya ximiya poverhnosti; *Mir: M.,* 568 (1979).
[48] Sharonov N. Yu.; Ulitin M. V. et al. Himiya i him. *Tehnologiya.* 52, 4, 11 (2009).
[49] *Adsorption from solution at the solid/liquid interface*; Editor, Parfitt G. D., Rochester C. H.; Academic Press Inc.: London, 488 (1983).
[50] *Magnetite: Structure, Properties and Application;* Editor, Angrove D. M.; Nova Science Publishers, Inc. 143-178, ISBN: 978-1-61761-839 (2011).
[51] Korolev V. V.; Ramazanova A. G. et al. *Colloid J.* 66, 700 (2004).
[52] Korolev V. V.; Blinov A. V. et al. *Colloid J.* 66, 704 (2004).

In: Oleic Acid: Dietary Sources, Functions and Health Benefits ISBN: 978-1-62618-332-2
Editor: Luciano Paulino Silva

Chapter 4

MICROBIAL TRANSFORMATIONS OF OLEIC ACID

Fabiano J. Contesini**, José V. Madeira Jr, Danielle B. Lopes, Paula Speranza, Elaine B. Ceresino, Camilo B. Teixeira and Gabriela A. Macedo

Department of Food Science, Faculty of Food Engineering, University of Campinas (UNICAMP), Campinas, Brazil

ABSTRACT

Oleic acid is one of the most important vegetable oils fatty acids within different industrial fields. This monounsaturated fatty acid is extensively found in olive, corn, and soybean oils, among others. Its importance varies from health-promoting properties, such as the prevention of cardiovascular diseases and some types of cancer, to many technological applications, taking into account the oxidative stability of this oil. One important application of oleic acid is its use for obtaining different compounds through microbial transformations and biocatalytic processes. These steps include hydration or epoxidation of the double bond, reduction of carboxylic acid and combinations of these reactions. An important group that can be obtained through transformation of oleic acid is the hydroxy fatty acids (HFA) and their derivatives. They are used in cosmetics, paints and coatings, lubricants and in the food industry. Several microorganisms, which include bacterial and yeast strains, produce HFA from oleic acid, such as in the obtainment of (E) 10-hydroxy-8-octadecenoic acid, using a *Pseudomonas* sp. 42A2 strain. 10-hydroxyoctadecanoic acid can be produced by *Nocardia cholesterolicum*, *Rhodococcus* sp. and *Saccharomyces cerevisiae*. These compounds are examples of monohydroxy derivatives of oleic acid and have a diverse potential for application in the food industry, considering its similarity with ricinoleic acid. Added to the possibility of biotransformation of oleic acid using microbial cells, the use of isolated lipases for the production of other compounds from oleic acid and/or triacylglycerols containing oleic acid have been reported, as well as the production of structured lipids by acydolysis of different oil sources with free fatty acid mixtures. These reactions are catalyzed by microbial lipases from different microorganisms like *Rhizomucor miehei*, and *Thermomyces lanuginosus*. Therefore, this review reports the use of free oleic acid and/or

* E-mail address: fabiano.contesini@gmail.com.

different oils containing oleic acid in their compositions for the obtaining of different important compounds, with the use of microbial cells by biotransformation of oleic acid, or by biocatalytic reactions using isolated microbial lipases.

INTRODUCTION

Oleic acid is a monounsaturated fatty acid, a member of the omega-9 series of fatty acids, which is found at high levels in soybean and olive oils. This lipid plays important roles in human life. Firstly, it presents interesting health properties, like the action of slowing the development of heart diseases. Moreover, oleic acid can be modified by chemical and enzymatic reactions to obtain important high value-added compounds. Considering that chemical methods use toxic reagents and generate a high number of byproducts, the microbial modifications of oleic acid seems to be much more attractive than these. Thus, the use of the enzymatic arsenal of microbial cells or their extracted enzymes, like lipases, for obtaining different compounds from oleic acid has been extensively studied.

The microbial conversion of fatty acids is a biological process that converts these lipids into new compounds with a large number of applications. Structural modifications via microbial bioconversion can alter the properties of lipids and even instil in them novel functions. Many economically valuable microbial products are secondary metabolites [1]. Microbial conversion employs fungus, yeast or bacteria to biocatalyze a specific step or steps in an otherwise strictly chemical synthesis [2, 3]. Microorganisms of industrial interest produce important compounds in high yields and grow rapidly in inexpensive culture media available in bulk quantities. These organisms are amenable to genetic manipulation and are nonpathogenic.

Microbial conversion of unsaturated fatty acids (for example oleic acid) has been widely exploited to produce industrial compounds, such as high value-added hydroxy products [4, 5]. Hydroxy fatty acids (HFAs) are known to exhibit special properties such as high viscosity and reactivity compared to normal fatty acids [6]. These modified fatty acids can be used in obtaining compounds of interest, such as resins, waxes, nylons, plastics, lubricants, cosmetics, and additives for coatings and paints. Additionally, HFAs are reported to contain antimicrobial activities against plant pathogenic fungi as well as several food-borne bacteria [7, 8, 9]. Studies report the bioconversion reactions by *Pseudomonas aeruginosa* strains that produce monohydroxy, dihydroxy and trihydroxy fatty acid derivatives from unsaturated fatty acids [10, 11, 12]. Biotransformation reactions of oleic acid catalyzed by bacteria species had been extensively reported [4, 10].

The application of isolated microbial lipases to biocatalyze oleic acid in a broad number of compounds is another important application of this lipid. Those compounds include biodegradable, biocompatible and environmentally-friendly biobased surfactants and emulsifiers widely employed in foods, pharmaceutics and cosmetics, such as saccharide-fatty acid esters [13] and propylene glycol mono- and diesters [14]. Other compounds can be obtained from the esterification of oleic acid using lipases, like phytosterol oleic esters [15] that have important health applications due to their greater capacity to lower plasma cholesterol concentration, compared to the free sterols. In addition, lipase-catalyzed interesterification reactions are important for obtaining structured lipids with important functional and technological properties.

This chapter discusses the microbial biotransformation of oleic acid in the obtaining of hydroxy fatty acids. In addition, different microbial lipase reactions of oleic acid to the production of high value-added lipids are reported on.

MICROBIAL BIOTRANSFORMATION OF OLEIC ACID

In microbial biotransformation, selected substrates can be converted into desirable products without decomposition and formation of unwanted byproducts. With mild reaction conditions, bioprocesses save not only operational energy, but also capital investment [16]. The bioconversion of fatty acids is, for this reason, a ''green'' technology that converts them into wholly new chemical compounds with antimicrobial, industrial or biomedical properties [17]. Vegetable oils are desirable, inexpensive feedstocks for the production of various bioproducts. Among them, soybean oil is an example of a relatively cheap raw material, which contains 22% oleic acid, and is an attractive consideration for bioindustries [18]. Olive oil is another rich source of oleic acid, with 78% content, but does not present as low a cost as soybean oil [19]. Microbial conversions of vegetable oil and their fatty acids have yielded several products that are potentially useful as value-added food and non-food items. This bioconversion can provide new products with enhanced functionality and reactivity. They are stereospecific and generate less undesirable by-products and residues [20].

Hydroxy fatty acids (HFAs), originally produced in nature mainly by plant systems, are known to exhibit special properties such as high viscosity and reactivity compared to normal fatty acids [21, 22]. Based on these structural peculiarities, HFAs possess high industrial potential in a wide range of applications, including resins, waxes, nylons, plastics, lubricants, cosmetics, and additives for coatings and paintings. In addition, HFAs are reported to contain antimicrobial activities against plant pathogenic fungi as well as several food-borne bacteria [23, 24]. Since HFA production is limited to plant systems, some researchers have focused on the production of important HFAs from various fatty acid substrates by microbial bioconversion [21]. The hydroxy fatty acids of microorganisms, which are produced by the biotransformation of unsaturated fatty acids, are classified according to the number of hydroxy groups such as mono- [25], di- [26], and trihydroxy fatty acids [27].

Recent researches have described compounds from the microbial conversion of unsaturated fatty acids as potential value-added products because they can inhibit or can be used in the biocontrol of certain important plant pathogens [28] and introduce new functional groups to the inner carbon atoms of fatty acids [18].

MAIN COMPOUNDS PRODUCED FROM OLEIC ACID BY MICROBIAL BIOTRANSFORMATION

Different novel products with enhanced functions of interest can be obtained from oleic acid by biotransformation. The product is dependent upon the microorganism used and environmental conditions, such as aeration and water content. Thus, there is an increasing interest in optimizing the production of these compounds in order to enable their use on a larger scale.

7,10 Dihydroxy-8(E)-Octadecenoic Acid (DOD)

DOD is a dihydroxy monoenoic C18 fatty acid uniquely carrying two hydroxy groups at carbons 7 and 10 and a *trans* double bond between carbons 8 and 9. HFAs like DOD have gained importance because of their special properties such as higher viscosity and reactivity compared with other non-HFAs [29].

Physiological activity tests of DOD revealed that this compound acts against *Bacillus subtilis* and a common yeast human pathogen, *Candida albicans*. DOD also inhibits the growth of many pathogenic bacteria including food-borne pathogenic bacteria such as *Escherichia coli* ATCC 8739, *Staphylococcus epidermidis* KCTC 1917, *Salmonella typhimurium* KCTC 2515, *Listeria monocytogenes* ATCC19111, *Staphylococcus aureus* ATCC 6538, and *Bacillus subtilis* ATCC 6501. It also inhibited the growth of the following plant pathogenic bacteria: *Pseudomonas syringae pv. sesami* KACC 10649; *Pseudomonas syringae pv. actinidae* KACC 10659; *Erwinia* sp. KACC 10207; *Pseudomonas syringae pv. syringae* KACC 10361; *Ralstonia solanacearum* KACC 10475; *Corynebacterium glutamicum* KACC 10784; *Pseudomonas corrugate* KACC 10141; *Xanthomonas axonopodis pv. citri* KACC 10443; *Xanthomonas campestris* KACC 10490; and *Clavibacter michiganesis* subsp. *michiganensis* KACC 20122 [18].

Chang et al. used triolein as a substrate for *Pseudomonas aeruginosa* (PR3) to produce DOD [29]. Triolein was first hydrolyzed into oleic acid by the triolein-induced lipase and then the released oleic acid was converted to DOD by PR3. Results from this study demonstrated that natural vegetable oils, without being intentionally hydrolyzed, could be used as efficient substrates for the microbial production of HFAs.

PR3 strains have been studied for their use in several unsaturated fatty acids in order to produce mono-, di- and tri-HFAs [22, 26, 30, 31, 32, 33]. This microbial culture was found to convert oleic acid to DOD very efficiently. The yield of DOD was originally at 60% [26] and was improved to over 80% yield through modification of culture conditions [32], and has now been improved to over 90% [33].

The production of DOD from oleic acid is unique since it involves both hydroxylation and possibly isomerization, an addition of two hydroxy groups at two positions and hence a possible rearrangement of the double bond of the substrate molecule [33]. Kuo and Lanser evaluated a scale-up conversion of oleic acid to DOD by P. *aeruginosa strain* PR3 (NRRL B-18602) in Fernbach flasks and bench-top bioreactors [34]. Among various factors being evaluated to affect the oleic acid bioconversion to produce DOD, the most significant findings were the need of pH adjustment prior to adding the substrate and the inability to increase yield by simply doubling the culture volume. The conclusion was that an effective bioreactor process could not be developed without an innovated aeration design to deal with the coexistence of a hydrophobic substrate and a surface-active product that is prone to foam formation.

Ellamar et al. studied the production of a novel biologically active furan fatty acid, the 7,10-epoxyoctadeca-7,9-dienoic acid (EODA), from DOD through a one-step heat treatment [35]. These lipids gain special attention due to the important roles that they play in biological systems, including humans. They are also known to have strong antioxidant activities such as radical scavenging activity. DOD produced from oleic acid by *Pseudomonas aeruginosa* PR3 was purified and converted in EODA by heat treatment in the presence of hexane as solvent.

The work provided useful information to produce a biologically active fatty acid that is cost-effective on a large scale.

10-HYDROXY-8(E)-OCTADECENOIC ACID (HOD)

HOD is an intermediate in the bioconversion of oleic acid to DOD (Figure 1). Studies indicate that this intermediate has applications as a substitute for ricinoleic acid [12-hydroxy-9(Z)-octadecenoic] (RA), which is the most important commercial source of HFA present exclusively in castor oil [22]. However, the fluctuating supplies and prices for castor oil resulted in a search for alternative raw materials, and finding substitutes for RA is of great interest.

Kuo et al. examined eighteen *Pseudomonas aeruginosa* strains for their ability to convert oleic acid to produce HOD as an effective alternative for producing RA or RA-like monohydroxy, monounsaturated fatty acids [36]. The results demonstrated that *P. Aeruginosa* strains possessed varying levels of activity for producing HOD. The ability to produce this compound was improved by optimizing such culture conditions as medium composition, reaction time, temperature, and pH. The authors concluded that the microbial production process might serve as an attractive source of HFAs, as an alternative to the castor bean, which contains ricin, a highly toxic protein, and other potent allergens, noxious to grow and process oil.

10-HYDROXYSTEARIC ACID (10-HSA)

The 10-HSA is optically active and has the D-configuration. The ability of cells to convert, under anaerobic conditions, oleic acid into 10-HSA was discovered almost half a century ago by Wallen et al. [37] in *Pseudomonas* sp. strain 3266 in 1962 with 14% yield.

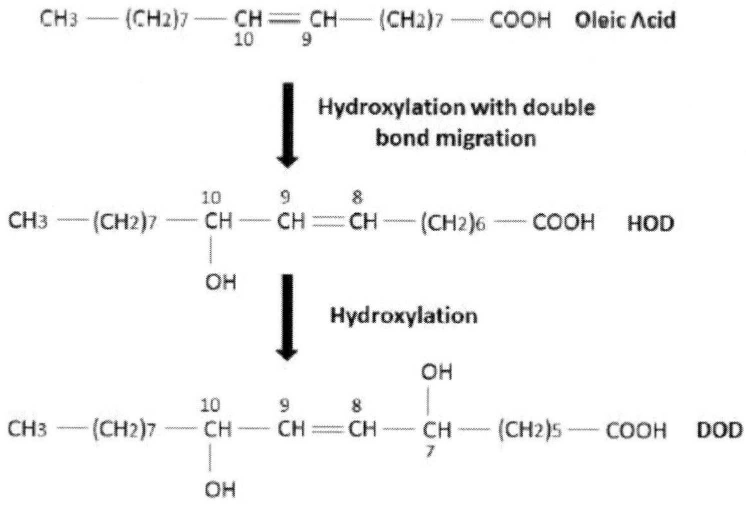

Figure 1. Conversion of oleic acid to HOD and DOD [adapted from 22].

In the following years, many other strains were also identified to be able to convert oleic acid into 10-HSA or to further oxidize it to 10-ketostearic acid (10-KSA) [38, 39, 40]. So far, microbial hydration of oleic acid was found in *Pseudomonas, Nocardia (Rhodococcus), Corynebacterium, Sphingobacterium, Micrococcus, Absidia, Aspergillus, Candida, Mycobacterium,* and *Schizosaccharomyces* [38, 41, 42].

The stereospecificity of 10-HSA varied depending on the microorganism. *Rhodococcus rhodochrous* ATCC 12674-mediated hydration of oleic acid gave mixtures of enantiomers 10-(R)-HSA and 10-(S)-HSA. *Pseudomonas* sp. NRRL B-3266 produced optically pure 10-(R)-HSA. *Sphingobacterium* produced 10-(R)-HSA containing 2–18% 10-(S)-HSA, and a *Flavobacterium* sp. DS5 produced 66% of enantiomeric excess of 10-(R)-HAS [16].

The conversion of oleic acid to 10-HSA by whole cells of *Stenotrophomonas nitritireducens* was optimized by Kim et al. [43]. Under the optimized conditions, the highest concentration, productivity, and yield of 10-HSA, with no byproduct, were obtained with a productivity of 7.9 $gL^{-1}h^{-1}$, the highest concentration reported thus far, contributing to the improved industrial production by means of a biological process.

10-Ketostearic Acid (10-KSA)

10-KSA is a major product of oleic acid bioconversion by a number of biological systems including a *Corynebacterium* sp. [44], one *Mycobacterium* and two *Nocardia* [38], a *Staphylococcus* sp. [45], and a *Flavobacterium* sp. strain DS5 [46]. The ability of microorganisms to convert oleic acid to either 10-KSA or 10-HSA as a main product may strongly be related to the cellular regulation of both hydratase and secondary alcohol dehydrogenase activities [46]. Data indicate that oleic acid is converted to 10-KSA via 10-HSA [16]. Production of 10-KSA was also reported by *Aspergillus terreus, Sphingobacterium thalpophilum,* and *Bacillus sphaericus* [47].

Heo et al. [19] used *Flavobacterium* sp. strain DS5 (NRRL B-14859) to convert olive oil and soybean oil to 10-KSA and 10-HSA. A small amount of lipase was added to the culture because the strain DS5 did not induce lipase activity to release free fatty acids from vegetable oils. The culture conditions were optimized, and 10-KSA production was higher from olive oil than from soybean oil due to the oleic acid content being present in greater quantities in olive oil.

Kuo and Levinson scaled up a bioprocess for the production of 10-HSA and 10-KSA, as well as their primary amides for potential new uses, using *Sphingobacterium thalpophilum* (NRRL B-14797) and *Bacillus sphaericus* (NRRL NRS-732), which solely produce 10-HSA and 10-KSA, respectively [48]. The scale-up reactor process provided crystalline 10-HSA and 10-KSA producing new primary amides via lipase-catalyzed amidation reaction with yields of 94 and 92%, respectively.

Enzymatic Reactions Using Oleic Acid As Substrate

Enzymes of lipid metabolism, isolated from microorganisms, are highly suitable for biocatalysts of fats and other lipids, considering most of these reactions do not require cofactors, and there is a large number of commercially available enzyme preparations.

Microbial lipases are the most important enzymes applied in lipids and as well as, and in particular, in oleic acid.

Most of the enzymes have high specificity, making the use of substrates for a particular enzyme limited. The oleic acid substrate presents this feature to fit very well for numerous isoforms of enzymes, mainly lipases [49]. Within this context, several types of reactions with oleic acid as substrates can be catalyzed by lipases. They include interesterifications, esterifications, hydrolysis, and epoxidation, to cite a few.

FUNCTIONAL LIPIDS

Phytosterols are known to have a hypocholesterolemic effect, allowing the reduction of low-density lipoprotein cholesterol in plasma, whereas high-density lipoprotein cholesterol concentration is not affected by their consumption [50]. Studies reported the use of lipases for esterification of phytosterols with oleic acid to decrease the melting point and increase the solubility of this compound, compared to free phytosterols.

Villeneuve et al. [50] evaluated the influence of various reaction parameters and the feasibility of esterification of canola phytosterols with oleic acid using the microbial lipases from *Rhizomucor miehei*, *Candida antarctica* B and *Candida rugosa*, observing that the nature of the acyl donor was of paramount importance with direct esterification with fatty acids, giving better results than transesterification with methyl esters. Similarly, the influence of substrates molar ratio was also checked, and a 1:3 excess of fatty acids was shown to be the best. Finally, others parameters such as optimal enzyme load (5%) or temperature (35°C) were determined to allow a final production of steryl esters close to 85% after 72 h.

Esteban et al. reported the synthesis of structured triglycerides of oleic acid in position sn-1,3 and palmitic acid in sn-2 position [51], which is interesting for nutrition, since it is naturally found in human milk. In the process, the authors used lipases from microbial lipase preparations in short-term reactions to incorporate oleic acids in the triacylglycerol simultaneously with the addition of palmitic acid in sn-2 position. Tecelão et al. reported the esterification of palmitic acid at the sn-2 position and oleic acid at sn-1,3 positions by commercially available immobilized lipases from *Rhizomucor miehei, Thermomyces lanuginose, Candida antarctica* and *Candida parapsilosis*, in a solvent-free process media [52]. These results are intriguing, considering that the structured lipids obtained is the same as the triacylglycerol presented in human milk.

SURFACTANTS/EMULSIFIERS

Monoglycerides (MGs) are anionic surfactants and emulsifying molecules with both hydrophilic and hydrophobic regions. They are widely used in food products, e.g. bakery products, margarines, dairy products and sauces [53]. In the cosmetic industry, they are texturing agents for improving the consistency of creams and lotions. The MGs are produced by chemical process, resulting in products with several drawbacks (dark color and burnt taste), and a low yield.

Alternatively, biocatalytic reactions catalyzed by lipases in non-aqueous media could be an alternative approach for MG synthesis, which could be performed by direct esterification of glycerol and fatty acids.

Other Applications

Marchetti and Errazu studied the influence of several variables to produce biodiesel [54]. The use of free fatty acid is an interesting alternative due to its low cost and is one of the promising substitutes for petroleum derivate fuels due to its environmental benefits. These authors studied different catalysts (resins, zeolite, and enzymes) to carry out the esterification reaction using different alcohols and oleic acid. The results showed a negative influence of alcohols carbon chain length and water on the conversion. Additionally, the lipase appeared to be the best choice, obtaining a yield of 98%.

Epoxidized oils are the most important products added to plastics, when we consider their greater flexibility and distensibility. The epoxidation of oleic acid is important as a plasticizer for polyvinyl chloride (PVC) and other plastics. These reactions can be catalyzed by lipase from *Candida antarctica* (Novozym® 435) as an alternative to the synthesis of epoxystearates, as shown in Figure 2 [55].

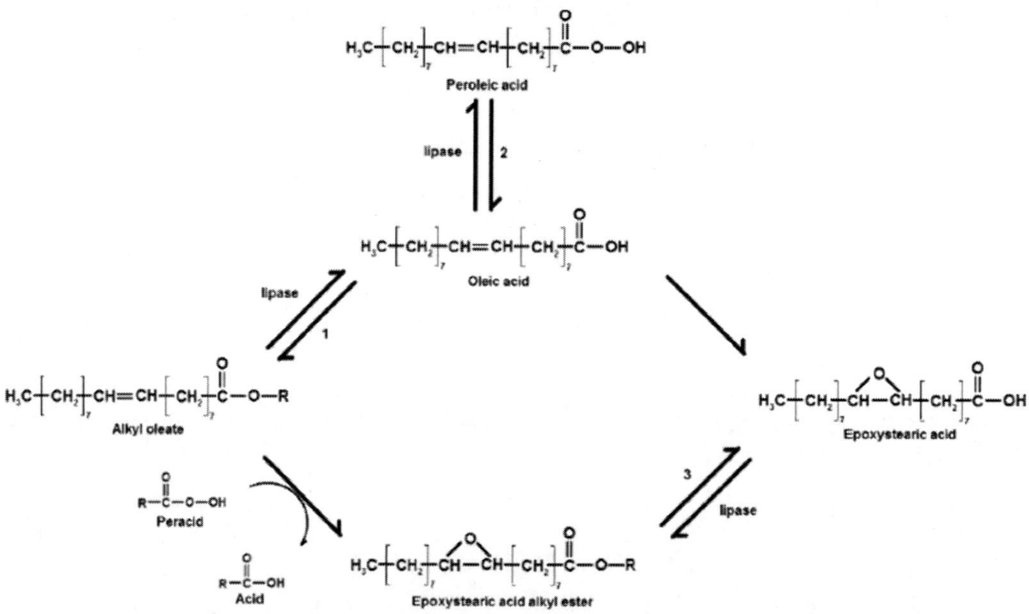

Figure 2. Representation of the reaction pathways for esterification catalysed by lipases: formation of 1 - Peracid; 2 - Oxygen is spontaneously transferred from the peracid to the double bond yielding epoxystearic acid; 3 - Epoxystearic acid can be used as a substrate for lipase catalysed esterification reaction [Adapted from 55].

Starch is a polysaccharide made up of a natural polymer, which is inexpensive, renewable, and fully biodegradable. Modification of starch hydroxy groups by esterification gives specific characteristics like thermoplasticity and water resistance to the starch ester over the unmodified starch. Unlike chemical esterification, enzymatic esterification is an environmentally friendly method which occurs under milder conditions [57]. The introduction of an ester group into polysaccharide constitutes an important achievement, considering it modifies their original hydrophilic nature for obtaining new thermal and mechanical properties [57].

Horchani et al. used a non-commercial $CaCO_3$-immobilized lipase from *Staphylococcus aureus* (SAL3) to catalyze the esterification between oleic acid and starch in pure substrate conditions using microwave heating followed by liquid state esterification obtaining in optimal conditions 76% of conversion during 4 h of incubation at 44°C [57]. Results showed that the α-amylolysis of the starch was significantly affected after esterification. The hydrophobic starch fatty acid esters produced may have potential industrial applications such as surface coating materials, flavoring agents in food industry and biomedical materials for bone fixation and replacements.

Akerman et al. reported the synthesis of trimethylolpropane (TMP) ester from oleic acid by immobilized lipase B from *Candida Antarctica* (Novozym®435) [58]. The TMP ester showed properties for use as hydraulic fluids (automotive applications), especially at extreme temperatures and other ecological properties for the biolubricants. It is easily biodegradable and possesses low aquatic toxicity.

In conclusion, the enzymatic process had a lower environmental impact and also yielded a product with interesting biolubricant properties in comparison to other heterogeneous catalysts.

Conclusions

Oleic acid is an important lipid with an unlimited number of applications. This is mainly attributed to the great possibilities of structural modifications of this monounsaturated fatty acid. Taking into account that the conventional chemical methods present many disadvantages, biotechnological tools are extensively focused to alternatively obtain oleic acid derivatives. The use of the enzymatic arsenal from bacterial cells, for example, has been shown to be a potential alternative to the obtainment of hydroxy fatty acids from oleic acid. These biotechnologically modified fatty acids specifically can be obtained due to the great enzymatic selectivity of the microbial enzymes. In addition to the use of a whole microbial cell to biotransform oleic acid, lipases are used to obtain different industrially important compounds, like surfactants, functional lipids, and biofuels, among others. It must be highlighted that biotransformation, using whole cells or isolated lipases, is considered a "green" process resulting in specific compounds with lower formation of byproducts.

REFERENCES

[1] J. Berdy, *J. Antibiot.* 58, 1 (2005).

[2] V. K. Bajpai, H. R. Kim, C. T. Hou and S. C. Kanget, *J. Ind. Microbiol. Biotechnol.* 36, 695 (2009).

[3] N. M. Carballeira, *Prog. Lipid Res.* 47, 50 (2008).

[4] C. T. Hou and M. O. Bagby, *J. Ind. Microbiol.* 7, 123 (1991).

[5] H. R. Kim, *Enzyme Microb. Technol.* 25, 583 (2001).

[6] M. O. Bagby and K. D. Calson, Fats for the future, Ellis Horwood Limited Press, Chichester, UK. (1989).

[7] V. Bajpai, S. Y. Shin, M. J. Kim, H. R. Kim and S. C. Kang, *Agric. Chem. Biotechnol.* 47, 199 (2004).

[8] C. T. Hou and R. J. Forman, J. *Ind. Microbiol. Biotechnol.* 24, 275(2000).

[9] S. Y. Shin, H. R. Kim and S. C. Kang, *Agric. Chem. Biotechnol.* 47, 205 (2004).

[10] H. Kim, H. W. Gardner and C. T. Hou, *J. Ind. Microbiol. Biotechnol.* 25, 199 (2000).

[11] T. M. Kuo, L. K. Manthey and C. T. Hou, *J. Am. Chem. Soc.* 75, 875 (1998).

[12] T. M. Kuo. H. Kim and C. T. Hou, *Curr. Microbiol.* 43, 198 (2001).

[13] R. Ye and D. G. Hayes, *J. Am. Oil Chem. Soc.* (2011) Doi:10.1007/s11746-011-1919-4.

[14] H. C. Chen, C. H. Kuo, W. C. Tsai, Y. L. Chung, W. D. Chiang, C. M. Chang, Y. C. Liu and C. J. Shieh, *J. Am. Oil Chem.* (2011) Doi: 10.1007/ s11746-011-1914-9.

[15] X. Pan, B. Chen, J. Wang, X. Zhang, B. Zhul and T. Tan, *Appl. Biochem. Biotechnol.* (2011) Doi: 10.1007/s12010-011-9324-x.

[16] C. T. Hou, Biocatalysis and Biotechnology for Functional Foods and Industrial Products, CRC Press, New York (2006).

[17] V. K. Bajpai, H. R. Kim, C. T. Hou, and S. C. Kang, *J. Ind. Microbiol. Biotechnol.* 36, 695 (2009).

[18] C. T. Hou, *Asia Pac. J. Clin. Nutr.* 17, 192 (2008).

[19] S. H. Heo, C. T. Hou and B. S. Kim, *New Biotechnol.* 26, 105 (2009).

[20] T. M. Kuo, T. Kaneshiro and C. T. Hou, *Lipid Biotechnology*, Marcel Dekker, New York (2002).

[21] J. H. Bae, M. J. Suh, N. Y. Lee, C. T. Hou and H. R. Kim, *Biotechnol. Bioproc. Eng.* 15, 953 (2010).

[22] H. Kim, H. W. Gardner and C. T. Hou, *J. Am. Oil Chem. Soc.* 77, 95 (2000).

[23] V. Bajpai, S. Y. Shin, M. J. Kim, H. R. Kim and S. C. Kang, *Agric. Chem. Biotechnol.* 47, 199 (2004).

[24] S. Y. Shin, H. R. Kim and S. C. Kang, *Agric. Chem. Biotechnol.* 47, 205 (2004).

[25] Y. Yamada, H. Uemura, H. Nakaya, K. Sakata, T. Takatori, M. Nagao, H. Iwase and K. Iwadate, *Biochem. Biophys. Res. Commun.* 226, 391 (1996).

[26] C. T. Hou, M. O. Bagby, R. D. Plattner and S. Koritala, *J. Am. Oil Chem. Soc.* 68, 99 (1991).

[27] T. M. Kuo, H. Kim and C. T. Hou, *Curr. Microbiol.* 43, 198 (2001).

[28] S. Y. Shin, V. K. Bajpai, H. R. Kim and S. C. Kang, *Int. J. Food Microbiol.* 113, 233 (2007).

[29] I. A. Chang, I. H. Kim, S. C. Kang, C. T. Hou and H. R. Kim, *Appl. Microbiol. Biotechnol.* 74, 301 (2007).

[30] C. T. Hou and M. O. Bagby, *J. Ind. Microbiol.* 9, 103 (1992).
[31] H. Kim, H. W. Gardner and C. T. Hou, *J. Ind. Microbiol. Biotechnol.* 25, 109 (2000).
[32] T. M. Kuo, L. K. Manthey and C. T. Hou, *J. Am. Oil Chem. Soc.* 75, 875 (1998).
[33] T. M. Kuo, H. Kim and C. T. Hou, *Curr. Microbiol.* 34, 198 (2001).
[34] T. M. Kuo and A. C. Lanser, *Curr. Microbiol.* 47, 186 (2003).
[35] J. B. Ellamar, K. S. Song and H. R. Kim, *J. Agric. Food Chem.* 59, 8175 (2011).
[36] T. M. Kuo, J. K. Huang, D. Labeda, L. Wen and G. Knothe, *Curr. Microbiol.* 57, 437 (2008).
[37] L. L. Wallen, R. G. Benedict and R. W. Jackson, *Arch. Biochem. Biophys.* 99, 249 (1962).
[38] S. H. El-Sharkawy, W. Yang, L. Dostal and J. P. Rosazza, *Appl. Environ. Microbiol.* 58, 2116 (1992).
[39] J. A. Hudson, C. A. M. MacKenzie, and K. N. Joblin, *Appl. Microbiol. Biotechnol.* 44, 1 (1995).
[40] T. Kaneshiro, L. K. Nakamura, and M. O. Bagby, *Curr. Microbiol.* 31, 62 (1995).
[41] L. E. Bevers, M. W. H. Pinkse, P. D. E. M. Verhaert and W. R. Hagen, *J. Bacteriol.* 191, 5010 (2009).
[42] A. Latrasse, S. Paitier, B. Lachot, P. Bonnarme, G. Féron, and A. Durand, *Biotechnol. Lett.* 19, 715 (1997).
[43] B. N. Kim, S. J. Yeom and D. K. Oh, *Biotechnol. Lett.* 33, 993 (2011).
[44] C. W. Seo, Y. Yamada, N. Takada and H. Okada, *Agric. Biol. Chem.* 45, 2025 (1981).
[45] A. C. Lanser, *J. Am. Oil Chem. Soc.* 70, 543 (1993).
[46] C. T. Hou, *Appl. Environ. Microbiol.* 60, 3760 (1994).
[47] S. Koritala and M. O. Bagby, *J. Am. Oil. Chem. Soc.* 69, 575 (1992).
[48] T. M. Kuo and W. E. Levinson, *J. Am. Oil. Chem. Soc.* 83, 671 (2006).
[49] T. M. Kuo, T. Kaneshiro and C. T. Hou, *Microbiological Conversions of Fatty Acids to Value-Added Products. Lipid Biotechnology.* T. M. Kuo, and H. W. Gardner Editors, New York, p. 605-628 (2002).
[50] P. Villeneuve, F. Turon, Y. Caro, R. Escoffier, B. Baréa, B. Barouh, R. Lago, G. Piombo and M. Pina, *Enzyme Microb. Technol.* 150, 37 (2005).
[51] L. Esteban, M. J. Jiménez, E. Hita, P. A. González, L. Martín and A. Robles, *Bioch. Eng. J.* 54, 62 (2011).
[52] C. Tecelão, J. Silva, E. Dubreucq, M. H. Ribeiro and S. Ferreira-Dias, *J. Mol. Catal. B: Enzym.* 65, 122 (2010).
[53] M. A. Jackson and J. W. King, *J. Am. Oil Chem. Soc.* 74, 353 (1997).
[54] J. M. Marchetti and A. F. Errazu, *Fuel* 87, 3447 (2008).
[55] C. Orellana-Coca, J. M. Billakanti, B. Mattiasson and R. Hatti-Kaul, *J. Mol. Catal. B: Enzym.* 44, 133 (2007).
[56] S. Sun, X. Ke, L. Cui, G. Yang, Y. Bi, F. Song and X. Xu, *Ind. Crops Prod.* 33, 676 (2011).
[57] H. Horchani, M. Chaâbouni, Y. Gargouri and A. Sayari, *Carbohyd. Polym.* 79, 466 (2010).
[58] C. O. Akerman, A. E. V. Hagström, M. A. Mollaahmad, S. Karlsson and R. Hatti-Kaul, *Proc. Biochem.* (2011) Doi: 10.1016/j.procbio. 2011. 08.006.

In: Oleic Acid: Dietary Sources, Functions and Health Benefits ISBN: 978-1-62618-332-2
Editor: Luciano Paulino Silva

Chapter 5

OLEIC ACID: CONVERSION OF FREE FATTY ACIDS INTO BIODIESEL OVER HETEROGENEOUS CATALYSTS

Catarina S. Caetano and José E. Castanheiro*
Centro de Química de Évora, Departamento de Química,
Universidade de Évora, Portugal

ABSTRACT

Biodiesel, which is recognised as a "green fuel", is an alternative to petroleum diesel due to its lower dependence on foreign crude oil and lower combustion emission profile (especially SO2). It is biodegradable and non-toxic, so it is beneficial from an environmental viewpoint. Biodiesel is generally synthesised by transesterification and esterification processes from vegetable oils, animal fats and wastes, which are carried out by acid or base catalysis.

The esterification reaction of long chain carboxylic acids, such as oleic acid, is interesting in the context of biodiesel production. In this process, acid catalysts are more efficient than base catalysts, which have led to the search for more active catalysts for both reactions.

Traditionally, these processes are carried out over homogenous catalysts. However, to make biodiesel production a "Green Process", homogenous catalysts have been replaced with heterogeneous ones. Heterogeneous catalysts have thus recently been used in esterification and transesterification reactions for biodiesel synthesis. Different solid acid catalysts, such as zeolites, resins, tungstated and sulphated zirconia, heteropolyacid, and ionic liquids, have been explored as heterogeneous catalysts for biodiesel production. In this chapter, the use of different solid catalysts for the conversion of free fatty acids (especially oleic acid) into biodiesel will be reviewed.

* E-mail address: jefc@uevora.pt

INTRODUCTION

Biodiesel is one attractive substitute for diesel fuel due to its lower dependence on foreign crude oil. It is also renewable and biodegradable. Furthermore, biofuel has limited greenhouse gas emissions because of the closed CO_2 cycle and lower combustion emission profile (especially SOx). It can be used without engine modifications and it has good engine performance. In fact, biofuel improves combustion because of its oxygen content. Finally, biodiesel has lower toxicity and can improve rural economies. The biofuel discussed here consists of fatty acid alkyl esters (usually methyl esters, FAME) derived from either the transesterification of triglycerides (TGs) or the esterification of free fatty acids (FFAs) with low molecular weight alcohols (Figure 1) [1-3]. Biodiesel is currently not cost competitive with conventional diesel fuel because of its high raw material and production costs. To improve the economical feasibility of biodiesel, cheap, low quality feed stocks such as waste cooking oil (WCO), animal fat and tall oil, which contain high free fatty acids (FFAs), have been used instead of refined vegetable oil. The amount of WCO generated in each country is huge and varies depending on their use of vegetable oil [4].

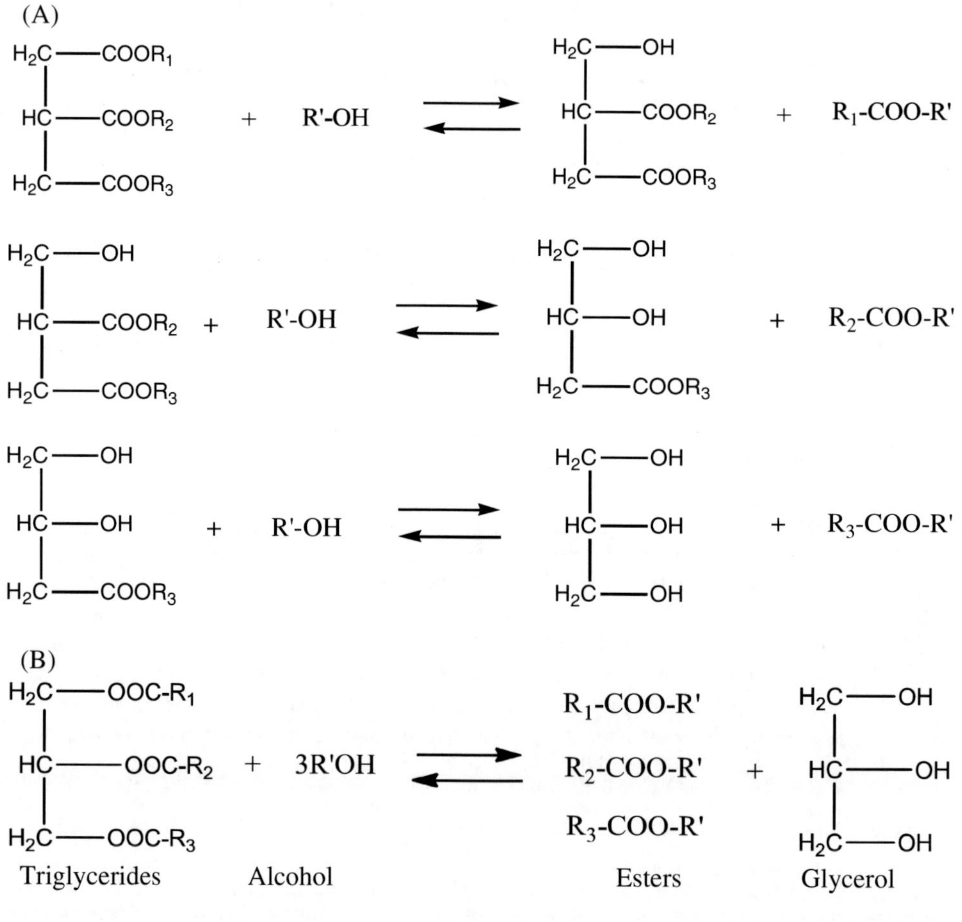

Figure 1. Transesterification reactions of triglycerides with alcohols. (A) Stepwise reaction; (B) Overall reaction.

Currently, the large quantity of inexpensive WCO from households and restaurants is collected and used either as animal feed or disposed, causing environmental pollution. Thus, WCO has significant potential as an alternative low cost biodiesel feedstock and partly decrease dependency on petroleum-based fuel.

The presence of FFAs is not compatible with homogeneous basic catalysts used in transesterification for biodiesel production. One way to overcome this problem is to subject the FFAs to esterification before the transesterification of triglycerides. Thus, esterification becomes an essential step in the production of biodiesel [5].

Traditionally, esterification and transesterification reactions are carried out over homogeneous catalysts, such as sulphuric acid or hydrochloric acid. However, these mineral acids are corrosive and have to be neutralised after reaction (with consequent salt formation). Therefore, to become a "Green Process", homogeneous catalysts must be replaced with heterogeneous ones. Solid acidic catalysts are preferable, offering easy separation from the reactants and products by filtration and allowing continuous operation of reactors. In addition, they do less harm to the environment and have no corrosion or disposal of effluent problems. Furthermore, they are easily separated from liquid products. Reusability of the catalyst is yet another advantage of heterogeneous catalysts.

As heterogeneous catalysts are insoluble in oil and methanol phases, they require high temperatures for optimum biodiesel yield. However, the application of heterogeneous catalysts for the industrial production of biodiesel requires minimal energy consumption. This scenario can be achieved if the heterogeneous catalysts are easily prepared and need only moderate reaction conditions. Hence, there is a need for the development of heterogeneous catalysts that can produce biodiesel at conditions (e.g., temperature and pressure) comparable to that used in homogeneous catalysis. The development of heterogeneous catalysts has been a relatively recent area of research in the synthesis of biodiesel [5, 6].

This chapter focuses on solid acid catalysts as potential heterogeneous catalysts for biodiesel synthesis. The first topic discussed is the esterification of fatty acids with alcohols (an homogeneous process). The second topic is the solid acid-catalysed esterification of free fatty acids, especially the esterification of oleic acid (an heterogeneous process). The third topic is the valorisation of waste cooking oil with high amounts of FFA (including oleic acid) into biodiesel over solid acid catalysts.

ESTERIFICATION OF FATTY ACIDS WITH ALCOHOLS (HOMOGENEOUS PROCESS)

To reduce the amount of FFA in the feedstock, it is necessary to introduce an intermediate step in the biodiesel production process, which is carried out over basic catalysts. One possibility for this step is the saponification of FFA with NaOH solution.

However, this process leads to partial consumption of the starting material and waste generation [7-8]. Another way to overcome this problem is to subject the FFAs to esterification with short-chain alcohols before the transesterification of triglycerides. Thus, FFA in the feedstock is also converted into biodiesel without waste generation. Therefore, esterification is an essential step in biodiesel production.

Traditionally, this reaction is carried out over homogenous catalysts. The esterification of FFA (e.g., oleic acid) can be catalysed by Brönsted acids. Figure 2 shows the mechanistic steps during the esterification of oleic acid with methanol.

The reaction is initiated by the catalyst, which donates a proton to the carboxylic acid. In the next step, the hydroxyl group of the alcohol (CH_3OH) attacks the carboxylic acid (nucleophilic attack). The reaction completes with water elimination.

The esterification of free fatty acids has been carried out over sulphuric acid [9-11] and alkyl sulfonic compounds [12, 13], which are most likely the most employed acid catalysts for this reaction. However, as a consequence of the application of these catalysts, high sulphur content is observed in biodiesel. In addition to their high corrosiveness, they are also not reusable and result in high generation of acid effluents and salts for disposal [14, 15]. For these reasons and many others, heterogeneous acid catalysts are preferred over homogeneous ones.

Esterification of Fatty Acids (Oleic Acid) with Alcohols in the Presence of Solid Acids

Esterification and transesterification have been largely applied to the production of biodiesel. The esterification reaction of long chain carboxylic acids, such as oleic acid, is interesting in the context of biodiesel production. This and other free fatty acids (FFA) may be present to different extents in transesterification reactions of vegetable oils and recycled feedstocks. As will be discussed, many different solid acids have been used as catalysts in the esterification of FFA with alcohols.

Resins with Sulfonic Acid Groups

Esterification of oleic acid with methanol has been carried out over cation-exchange membranes at room temperature. Use of ion-exchange membranes involves liquid phases, prevents the appearance of the compensating ion of the proton in the reaction medium and allows extraction of water from the reaction mixture.

High conversion, generally greater than 80%, is reached after 8 h. Using ion-exchange membranes in a stack reactor allows exclusion of the compensating anions of the catalysing protons from the reacting mixture, and as a consequence avoids their elimination with special down-stream treatments. Due to the equilibrium established through the cation-exchange membrane, protons are kept in a static layer on the membrane in the reacting mixture. As a consequence, proton content is constant and can be raised to a high level. Furthermore, this constant level avoids any regeneration of the catalyst [16,17].

The production of decanoic acid methyl esters by esterification of the fatty acid decanoic acid with methanol was carried out on a strong acidic ion-exchange resin (Amberlyst 15). A kinetic model based on a Langmuir-Hinshelwood-Hougen-Watson approach was derived and the kinetic constants of this and a pseudo-homogeneous model were fitted [18].

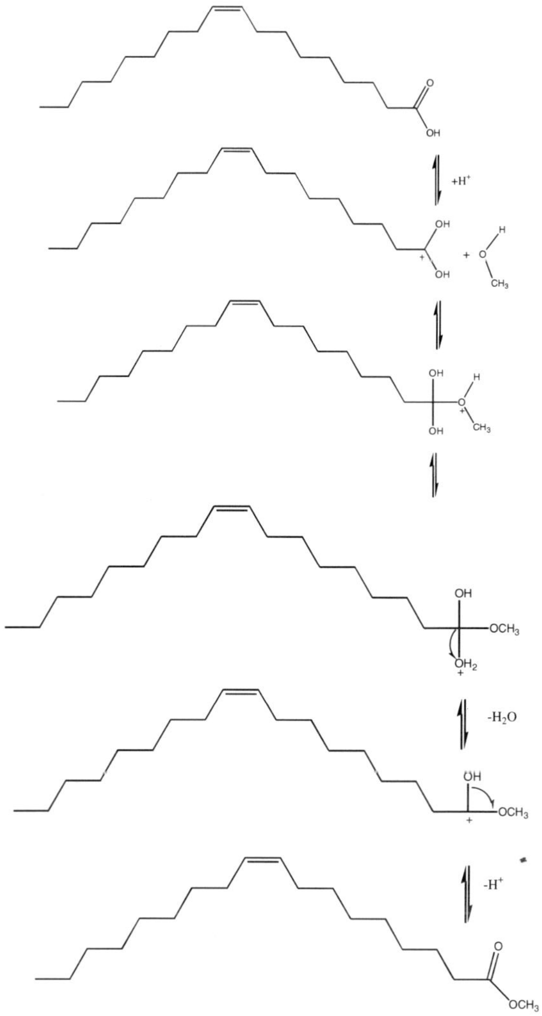

Figure 2. Mechanism of acid catalyzed esterification of oleic acid with methanol.

The production of biodiesel from free fatty acids (FFAs) contained in by-products of a vegetable oil refinery was studied using the commercial super acidic resin Purolite CT-275. The experiments were performed in a batch reactor in the temperature range of 363.15–393.15 K with an initial molar ratio of methanol to FFAs of 6.6/1, while the catalyst mass was fixed at 2 wt% of the total vegetable oil mass. A kinetic model was developed that accounted for the reversible esterification reaction. The esterification kinetic model developed excellently predicted the esterification extent of two other oils with much higher FFA content [19]. Biodiesel synthesis via fatty acids esterification with methanol was carried out over cation-exchange resin (Dowex monosphere 88). The reaction rate and fatty acid conversion increased with increasing catalyst loading, reaction temperature and molar ratio of feeding reactants. A pseudo-homogeneous kinetic model coupling the effect of catalyst loading, reaction temperature and methanol/fatty acids molar ratio gave a correlation coefficient of 0.95 between experimental and predicted data [20].

Esterification of oleic acid with ethanol was carried out over sulphonated polystyrene. This catalyst, with acidic and superabsorbent properties, was obtained by sulphonation of expanded polystyrene. It showed a high concentration of active sulphonic acid sites (0.7–5.9 mmol acid sites g^{-1}) and high water absorption capacity (445–900 $g_{water}.g^{-1}$). It was observed that the catalytic activity increased with the number of acid sites and water absorption capacity. Turnover frequencies suggested that the catalytic activity depended on the accessibility/diffusion determined by the degree of crosslinking in the polymer. These catalysts could be removed from the medium by simple decantation or filtration and reused several times [21].

Petchmala et al. studied the esterification of oleic acid with ethanol over expanded polystyrene with sulphonic acid sites (SAP) [22]. The authors observed that the catalytic activity increased with the number of acid sites and the water absorption capacity. Turnover frequencies suggested that the catalytic activity depended on the accessibility/diffusion determined by the crosslink in the polymer. Commercial sulphonic acid resins and polyacrylate-based superabsorbent polymers showed very low activities compared with the SAP produced. The SAP also showed higher activity compared to the homogeneous catalyst p-toluenesulphonic acid. These SAP catalysts can be removed from the medium by simple decantation and filtration and reused several times.

Esterification of oleic acid was also studied in a three-phase fixed-bed reactor with a cation exchange resin catalyst (Amberlyst-15) at high temperature (80 to 120ºC). The fatty acid methyl ester (FAME) yields in the fixed-bed reactor were increased with increasing reaction temperature, methanol flow rate and bed height. Moreover, the FAME yields were higher than those obtained using a batch reactor due to an equilibrium shift toward the product that resulted from continuous evaporation of the produced water. In addition, there was no catalyst deactivation during the esterification of oleic acid in this report [23].

Caetano et al. reported the esterification of palmitic acid with methanol using poly(vinyl alcohol) cross-linked with sulphosuccinic acid (SSA) and polystyrene cross-linked with divinylbenzene containing sulphonic acid groups as catalysts at 60ºC [24]. It was observed that the PVA matrixes with sulphonic acid groups were more active than the polystyrene ones. With the most active polymeric matrix (PVA_SSA40), an almost 90% conversion was achieved after 2 h. The influence of various reaction parameters, such as molar ratio of palmitic acid to alcohol, catalyst loading and type of alcohol, on the activity of the most efficient catalyst, PVA_SSA40, was studied. In relation to the molar ratio of palmitic acid to alcohol, it was observed that when the molar ratio increased, the equilibrium conversion of palmitic acid increased from approximately 30% (1:3) to 90% (1:63). The esterification of palmitic acid with ethanol led to lower conversion. When the temperature was increased from 60ºC to 80ºC, an increase in the palmitic acid conversion was observed using ethanol.

Catalytic stability of the PVA_SSA40 was evaluated by performing consecutive batch runs with the same catalyst sample. After the second batch, a stabilisation of the catalytic activity was observed. PVA_SSA40 was also used as a catalyst in the esterification of oleic and stearic acid with methanol, leading to an equilibrium conversion of almost 80%.

Esterification of oleic acid with methanol was carried out in the presence of Amberlyst-15 as a heterogeneous catalyst or of sulphuric acid as a homogeneous catalyst. The yield of fatty acid methyl ester (FAME) was studied at oil to methanol molar ratios of 1:3 and 1:6 and reaction temperatures of 60 and 80ºC. The rate of esterification of oleic acid significantly decreased as the initial water content increased to 20% of the oil. The activity of Amberlyst-

15 decreased more rapidly than that of sulphuric acid, due to the direct poisoning of acid sites by water. The presence of water thus resulted in poor accessibility of reactants to the acid sites. The reported two-step esterification of oleic acid, which included replacing the methanol and catalyst with fresh methanol and catalyst after 30 min, reduced the reaction time required to reach FAME of over 90% [25].

Morales et al. studied the esterification of oleic acid with n-butanol over organosulphonyl -modified mesostructured silica materials as catalyst [26]. Organically modified mesoporous silica materials were prepared by direct co-condensation of styrylethyl-trimethoxysilane (STETMOS) and tetraethyl-orthosilicate (TEOS) in a one-pot synthesis. The resultant organosulphonic-modified mesostructured silica materials exhibit hexagonal long-range mesoscopic arrangement with extended surface areas and narrow mean pore size distributions. Upon sulphonation, a high number of sulphonic-acid sites were introduced into the silica-anchored polystyrene-type organic moieties, thus providing strong acid sites embedded in a hydrophobic microenvironment. It was observed that these catalysts had high catalytic activity, especially in hydrophobicly demanding systems.

Sulfated Metal Oxides and Mixed Metal Oxides

Sulphated tin and zirconium oxides and tungstated zirconia were used in the transesterification of soybean oil with methanol at 200–300°C and the esterification of n-octanoic acid with methanol at 175–200°C. Tungstated zirconia–alumina is a promising solid acid catalyst for the production of biodiesel fuels from soybean oil because of its high performance and conversions over 90% for both of the described esterifications [27].

Esterification of dodecanoic acid with 2-ethylhexanol has been carried out over sulphated metal oxides (zirconia, titania and tin oxide). The sulphated zirconia was found to be a good candidate as it was active, selective, and stable under the process conditions. This made sulphated zirconia the most appealing candidate for catalytic biodiesel production [28].

Ropero-Vega et al. used sulphated titanias as catalysts in the esterification of oleic acid with methanol [29]. The catalysts were prepared using ammonium sulphate and sulphuric acid as sulphate precursors. Depending on the sulphation method, important effects on the acidity, textural properties and activity were found. The amount and method used for adding the sulphate ions determined the strength of the acidic properties. The sulphated titanias showed very high activity for the esterification of fatty acids with ethanol in a mixture of oleic acid (79%). Conversions up to 82.2% for oleic acid and a selectivity to ester of 100% were reached after 3 h of reaction at 80°C.

Modified zirconias (titania zirconia, sulphated zirconia, and tungstated zirconia) have been evaluated for both esterification and transesterification. The esterification of free fatty acids (caprylic acid) with ethanol was carried out over tungstated zirconia. The activity results for WZ indicated how biodiesel could be efficiently produced from free fatty acid-containing feedstocks using a solid acid-catalysed pre-esterification step, and how a significant fraction of the triglycerides could also be converted at the same time through initial hydrolysis using the water by-product from the esterification [30].

The esterification of oleic acid by n-butanol has been studied over silica-supported zirconium sulphate (ZS). These catalysts were prepared by impregnation with 10–50 wt% loading. It was found that most of the ZS in the samples was well dispersed on the support.

However, at loadings higher than 40 wt%, ZS existed as small agglomerated crystalline particles with an average size of 23.8–32.1 nm. The catalytic activity of silica-supported ZS was found to be higher than that of bulk ZS and even higher than that of conventional acidic resins. Different ZS loadings also had a strong impact on the structure and reactivity of silica-supported ZS catalysts. It was observed that the catalytic activity increased at loadings up to 30 wt%, while a decrease was found at higher loadings due to a small amount of agglomeration of ZS at higher loadings. The reusability of silica-supported ZS was much better than that of bulk ZS. It was observed that esterification of fatty acid was possible at short reaction periods and low catalyst loading under solvent-free conditions to achieve a greener reaction [31].

The esterification of oleic acid with n-butanol was also carried out over zirconium sulphate (ZS) supported on a pure hexagonal mesoporous material (HMS). Interestingly, no ZS crystal phase was developed even at ZS loadings as high as 40 wt%. The interaction of ZS with the HMS material enhanced the acidity and the catalytic activity of this reaction. The HMS-supported ZS catalyst exhibited higher catalytic activity than bulk ZS esterification of oleic acid with n-butanol under solvent-free conditions. The optimal loading of ZS was found to be 40 wt%. Most ZS samples are well dispersed on the support, except for 50 wt% loading, where the sample existed as small agglomerated particles with an average size of 30.7 nm.

Although some structure distortion was observed after loading with ZS, the activity of HMS-supported ZS was higher than that of bulk ZS. Optimal activity, reusability and thermal stability was obtained for the HMS-supported ZS catalyst with 40 wt% loading. Kinetic studies have shown that the esterification of oleic acid with n-butanol obeys an ER mechanism. It was noteworthy that esterification of fatty acid could be carried out at shorter reaction periods and with smaller amounts of catalyst under solvent-free conditions to make them greener reactions [32].

Recently, Moreno et al. reported the esterification of oleic acid with ethanol over sulphated tin oxides (SO_4^{2-}/SnO_2) [33]. Catalysts with different SO_4^{2-} contents (0.15, 0.3 and 0.45 wt%) were synthesised from hydroxylated tin oxide obtained by precipitation, followed by ion exchange of the hydroxyl groups with SO_4^{2-} species using a sulphuric acid solution. Subsequent stabilisation of crystallite size by the SO_4 groups on the SnO_2 surface was clearly observed. Sulphated tin oxides were tested in the esterification of free fatty acids with ethanol (molar ratio 1:10) at 353 K. The conversion of oleic acid increased with increasing surface acidity, reaching a maximum when the sulphate content and calcination temperature were 0.3 wt% and 773 K, respectively. Regeneration through heating the SO_4^{2-}/SnO_2 for 4 h allowed it to be reused without losing catalytic activity.

Esterification of oleic acid with ethanol to produce ethyl oleate, which is an important constituent of biodiesel, was also carried out on a solid acid catalyst of mesoporous SnO_2/WO_3. The synthesised mesoporous SnO_2/WO_3 exhibited 90% conversion of oleic acid and the corresponding yield of ethyl oleate was 92%. Kinetic investigation revealed that the experimental data followed a first order dependency on the concentration of the oleic acid and the catalyst. The conversion of oleic acid also depended on the reaction temperature and alcohol employed. The catalyst was recovered after the catalytic test, regenerated by heating at 400ºC for 3 h and successfully reused in the same esterification reaction [34].

Another material that has been used as a catalyst in the esterification of oleic acid with methanol at 65ºC is zirconium-doped MCM-41 silica-supported WO_x. A series of zirconium-doped MCM-41 silica supported WO_x solid acid catalysts with WO_3 loading ranging from 5 to 25 wt% were prepared by impregnation with ammonium metatungstate and subsequent activation at temperatures between 450 and 800°C. WO_3 loadings of 15–20 wt%, after activation at 700°C, led to the most active catalysts with conversions close to 100%. The catalyst with 15 wt% WO_3 was stable even when the reaction was carried out at 200°C, and reusable for at least four cycles with no leaching of tungsten [35].

More recently, Costa et al. reported the esterification of oleic acid with ethanol over WO_3 supported on a USY zeolite [36]. These catalysts were prepared by wet aqueous impregnation of tungstic acid on HUSY with loadings from 2.5 to 20.5 wt% WO_3. A high dispersion of WO_3 species on USY was subsequently achieved. A decrease in the micropore volume confirmed that tungsten species were preferentially deposited inside the zeolite structure. USY had the most and the strongest acidic sites (Brønsted type), but the incorporation of WO_3 decreased the amount and the strength of these new sites. However, all WO_3/USY catalysts were more active than USY in the esterification of oleic acid with ethanol (conversion above 74%, 2 h at 200°C). The calculation of the TOF for a 1 h reaction demonstrated that 11.4% WO_3/USY was the most active catalyst. Deactivation studies showed a direct correlation with reduction of the number of acid sites after the reaction and decreased activity upon recycling. The better performance of the 11.4% WO_3/USY sample could also be attributed to a better distribution in the strength of the acidic sites and an increased hydrophobic character of the synthesised material. Moreover, this material showed better resistance to water deactivation and thus had a lower decrease in conversion for recycling.

Heteropolyacids

Heteropolyacids (HPAs) are typically strong Brønsted acids that catalyse a wide range of reactions. The major disadvantages of HPAs as catalysts are their low thermal stability, low surface area (1-10 m^2/g), difficulty in separation from reaction mixtures and poor solubility. To increase the specific area of heteropolyacids, a variety of supports like zeolite, activated carbon, silica and polymers have been used to immobilise HPAs [37, 38].

Esterification of oleic acid with ethanol was carried out over $H_3PW_{12}O_{40}$ supported on zirconia (ZrO_2). Catalysts with different heteropolyacids in different ratios (5, 10, 15, 20, 25, 40 and 60 wt%) were prepared for this study. The samples were subsequently calcined at 200°C for 4 h. No decomposition of the Keggin anion structure was observed under these conditions. Optimum reaction performance (88% oleic acid conversion) was achieved at 20 wt% loading, 100°C, 4 h reaction and a 1:6 (acid:ethanol) molar ratio. This catalyst showed a minor leaching of 8 wt% (related to the initial loading), which affected the reaction kinetics. Preliminary tests of recyclability for 20% H_3PW/ZrO_2 indicated that treatment involving a sequence of washing with n-hexane, drying at 100°C and calcining at 300°C for 4 h recovered the conversion to 70% (80% of the original value) [39].

Narasimharao et al. reported the esterification of palmitic acid and transesterification of tributyrin, important steps in fatty acid and ester processing for biodiesel synthesis, over insoluble HPA salts of general formula $Cs_xH_{3-x}PW_{12}O_{40}$ ($x = 0.9$–3.0) [40]. It was observed that optimum performance was achieved with Cs loadings of $x = 2.0$–2.3. All samples with $x > 1$ were resistant to leaching and could be recycled without a major loss of activity.

$Cs_{2.3}H_{0.7}PW_{12}O_{40}$ could also be used in simultaneous esterification and transesterification without loss of activity or selectivity.

Esterification of palmitic acid with methanol was carried out over tungstophosphoric acid (PW), molibdophosphoric acid (PMo) and tungstosilicic acid (SiW) immobilised on silica by a sol–gel technique at 60ºC. It was observed that the catalytic activity decreases in the order: PW-silica > SiW-silica > PMo-silica. A series of PW catalysts immobilised on silica with PW loadings from 2.5 wt% to 8.4 wt% was prepared. It was observed that the PW-silica2 (4.2 wt%) showed the highest catalytic activity in the esterification of palmitic acid with methanol. Catalytic stability of the PW-silica2 was evaluated by performing consecutive batch runs with the same catalyst sample. After the second batch a stabilisation of the initial activity was observed [41].

To study heteropolyacid lixiviation, a posterior catalytic experiment was carried out. The PW-silica2 was put in contact with methanol, without palmitic acid, for 72 h. After this period, the catalyst was separated from methanol by centrifugation and palmitic acid was added to the reaction mixture. It was observed that the reaction only occurred in the heterogeneous phase. The PW-silica2 was also used as a catalyst in the esterification of oleic and stearic acid. Good catalytic activity was also observed for these substrates over PW-silica2 [41].

Alsalme et al. reported the esterification of hexanoic acid with methanol (1:20 molar ratio) over a range of HPA catalysts, both homogeneous and heterogeneous, in comparison to conventional homogeneous and solid acid catalysts (e.g., H_2SO_4, Amberlyst-15 and zeolites HY and H-Beta) [42]. The intrinsic catalytic activity (turnover frequency, TOF) of HPA catalysts was significantly higher than that of the conventional acid catalysts in these reactions. The TOF values decreased with decreasing catalyst acid concentration. However, supported HPA catalysts suffered from leaching and exhibit significant homogeneous catalysis by the leached HPA.

$H_3PW_{12}O_{40}/Ta_2O_5$ composite catalysts prepared by the sol–gel-hydrothermal method in the presence of triblock copolymer surfactant were used in the esterification of lauric acid with ethanol, an important pretreatment step in the production of biodiesel from renewable feedstocks. The catalysts, with controllable $H_3PW_{12}O_{40}$ loadings and well-defined mesoporous structures, showed high activity, selectivity and stability [43].

A series of catalysts containing 5–30 wt% of 12-tungstophosphoric acid (TPA) supported on niobia were prepared and their catalytic activity was studied in the esterification of free fatty acids (palmitic acid) with methanol. The catalytic activity depended on the content of TPA on the niobia. The authors also observed that the catalyst with 25 wt% TPA exhibited the highest activity. The esterification activity was dependent on the acidity of the catalyst, which was dependent on the presence of intact Keggin ions in the support [44].

Recently, Tropecêlo et al. reported the esterification of palmitic acid with methanol at 60ºC using different heteropolyacids (tungstophosphoric acid (PW), molybdophosphoric acid (PMo) and tungstosilicic acid (SiW)) immobilised on SBA-15 [45]. All of these catalysts exhibited high catalytic activity in palmitic acid esterification with methanol. It was observed that the catalytic activity decreased in the order: PW1-SBA-15 > SiW-SBA-15 > PMo-SBA-15. A series of PW catalysts immobilised on SBA-15 with PW loadings from 2.7 wt% to 8.3 wt% was subsequently prepared. High catalytic activity was observed with low tungstophosphoric acid loadings immobilised on the SBA-15. To optimise the reaction conditions, the effect of catalyst loading, carbon length of the alcohol, temperature, and molar ratio of fatty acid to methanol in the presence of PW3-SBA-15 were studied. In addition, the PW3-SBA-15 catalyst could be separated from the reaction system for reuse. A small amount of leaching of the heteropolyacid from SBA-15 to the liquid phase was observed. Due to the high catalytic activity of PW3-SBA-15 on the esterification of palmitic acid with methanol, the catalyst was also tested in the esterification of stearic and oleic acid with methanol. The catalytic activity was observed to decrease with an increasing number of carbon atoms.

Biodiesel production by esterification of oleic acid with methanol was carried out in the presence of 12-tungstophosphoric acid anchored to SBA-15. The influence of reaction parameters (such as catalyst concentration, acid/alcohol molar ratio and reaction temperature) on catalytic performance was then studied. The catalyst exhibited high activity for the esterification of oleic acid with methanol under mild conditions. It was also observed that the turnover frequency was 9.3 min^{-1}. This catalyst also had the potential for being a recyclable catalyst material after simple regeneration. Kinetic investigation revealed that the esterification of oleic acid with methanol followed a first order dependency on the concentrations of oleic acid and catalyst [46].

12-tungstophosphoric acid anchored to MCM-41 was used as a catalyst in biodiesel production via esterification of palmitic acid under relatively mild conditions. The catalyst exhibited high catalytic activity, which was mainly attributed to the large surface area and pore diameter of the mesoporous support (MCM-41) and the Brønsted acid strength of the TPA active sites. In addition, the catalyst showed the potential of being used as a recyclable catalytic material after a simple regeneration without any significant loss in conversion. Kinetic studies showed that the esterification of palmitic acid followed first order kinetics. [47]. Even more recently, Srilatha et al. reported the esterification of palmitic acid with methanol over 12-tungstophosphoric acid (TPA) supported on a SnO_2 as a catalyst [48]. The esterification activity depended on the amount and dispersion of TPA on the SnO_2, which in turn depended on the acidity. Among the catalysts, 15 wt% TPA/SnO_2 was the most promising one with the highest fatty acid conversion. The effect of various parameters on the reaction was studied and the optimum conditions found were a stirring speed of 600 rpm, a catalyst concentration of 0.055 $g.cm^{-3}$, a methanol to acid mole ratio of 14:1 and a reaction temperature of 65°C. The stability of the catalyst was confirmed by observing no leaching of TPA from the support. Studies to assess the catalyst reusability implied that its activity was consistent for 5 reaction cycles. A pseudo first order model was applied to correlate the experimental kinetic data and the kinetic parameters. The activation energy of the catalysts was comparable to that of mineral acids and lower than typical solid acid catalysts.

Zeolites

Zeolites are active catalysts for the esterification of large carboxylic acids, but they catalyse this reaction rather slowly. Thus, only large-pore zeolites have been used with any success in fatty acid esterifications. Even in these successful cases, however, the reaction always gave a variety of undesired by-products due to the high reaction temperature used. In general, the catalytic activity of zeolites for the esterification of fatty acids increases with increasing Si/Al ratio, indicating that reactivity was influenced by acid site strength and surface hydrophobicity. In conclusion, pore size, dimensionality of the channel system (related to the diffusion of reagents and products), and aluminium content of the zeolite framework strongly affected zeolite catalytic esterification activity [49]. As an example, the esterification of dodecanoic acid with 2-ethylhexanol was carried out over zeolites (Beta, ZSM-5, Y). These zeolites showed only a small increase in conversion (1 – 4%) compared to the non-catalysed reaction. This suggested that the reaction might have been suppressed by the limited diffusion of the bulky reactant into the zeolite pores. Therefore, catalysis most likely only took place on the external surface [50].

Activated Carbons

Carbon-based solid acids are considered ideal catalysts for many reactions due to their chemical inertness and superior mechanical and thermal stability. Sulphonated carbon is among the most promising of the solid acids developed for FFA [51, 52]. Recently, a new class of sulphonated carbons derived from incomplete carbonisation of simple natural products, such as sugar, starch or cellulose, has been reported to have better catalytic performance for the esterification of fatty acids and higher stability than sulphonated mesoporous silicas [53, 54].

Mesoporous carbon, which can be replicated from mesoporous silica and whose surface is hydrophobic, can be an ideal catalyst for the esterification of fatty acids. The acid-catalysed esterification of oleic acid with methanol was carried out over mesoporous carbon with sulfonic acid group functionalisation.

The catalysts were prepared through first sulphonating the carbon/SBA-15 hybrid composite prior to silica removal. It was observed that the sulphonation before silica removal was to maintain the ordering of the porous structure. Soft carbonisation was important to introduce enough acidic sites. The optimal mesoporous SO_3H–carbon material (CMK-3-873-SO_3H) exhibited a high surface area and a uniform pore size. The large pore size and hydrophobic surface of CMK-3-873-SO_3H made it a highly efficient esterification catalyst. The catalyst CMK- 3-873-SO_3H was also reusable; a high activity was observed even after four recycles. The increased catalytic performance was attributed to the hydrophobic surface and larger pore size of the new catalyst, which could effectively accommodate long chain fatty acids and reject formed water [55].

Recently, Geng et al. reported the esterification of oleic acid with methanol over a series of carbon-based solid acid catalysts. The catalysts were prepared by the sulphonation of mesoporous carbon substrates with thin pore walls [56]. Such catalysts were composed of thin, small carbon sheets, which facilitated their dispersion in polar solvent. The highest turnover frequency (TOF) observed was 78 h^{-1}, five times that of Amberlyst-15. The high catalytic activity was possibly attributed to the good dispersion of the catalysts in methanol. Catalysts with improved dispersion were obtained by a modified preparation, after which the highest TOF observed was 109 h−1, seven times that of Amberlyst-15.

Carbon-based solid acid catalysts prepared by the sulphonation of carbonised vegetable oil asphalt and petroleum asphalt showed catalytic activity in biodiesel production from model waste oil (with a high amount of oleic acid). The vegetable oil asphalt-based catalyst showed the highest catalytic activity.

The high activity was ascribed to the high density and stability of its acid sites, loose irregular network, and the hydrophobicity of its carbon sheets that prevented the hydration of –OH groups when water was present. In addition, large pores provided more acid sites to the reactants. These materials could simultaneously catalyse transesterification and esterification, so they should prove to be useful for the conversion of waste oils with high FFA concentrations into biodiesel [57].

Acidic Ionic Liquid

Esterification of fatty acids with methanol was carried out over a modified bentonite with 1-benzyl-1H-benzimidazole-based Brønsted acidic ionic liquids as catalysts. Three bentonites modified with acidic ionic liquids showed good catalytic activity and reusability for the esterification of oleic acid with methanol. It was observed that MB3B (bentonite modified with 3,30-(butane-1,6-diyl)bis(6-sulpho-1-(4-sulphobenzyl)-1H-benzimidazolium) hydrogen-sulphate) had the highest catalytic activity and best recyclability under the optimised reaction conditions. To further optimise the reaction conditions, the effect of reaction temperature, type and amount of catalyst, molar ratio and reaction time was investigated. A conversion of more than 92% was obtained for the esterification of oleic acid with methanol by varying the reaction parameters, including temperature, molar ratio, amount of catalyst and reaction time. Furthermore, high yields (85.5–98.3%) were obtained for the esterification of various carboxylic acids and alcohols under the new optimised reaction conditions [58].

Other Catalysts

The esterification of oleic acid in subcritical methanol was carried out over zinc acetate in a batch-type autoclave. The results suggested that zinc acetate was catalytically active in the esterification of oleic acid. Because all the components could be dissolved in subcritical methanol, the esterification was promoted under subcritical conditions. The oleic acid conversion reached 95.0% at 220ºC and 6.0 MPa with the molar ratio of methanol to oleic acid of 4:1 and 1.0 wt% zinc acetate as catalyst [59].

Recently, Nascimento et al. reported the esterification of free fatty acids (oleic acid) with methanol over two Amazon kaolins (Century and flint) and two standard kaolins (KGa-1b and KGa-2) that were thermally treated at 950°C and leached with 4 M sulphuric acid [60]. All of the leached metakaolins studied showed catalytic activity for the esterification of oleic acid with methanol. The analysis of conversion values obtained revealed a correlation with aluminium content and acidity for each prepared sample. In fact, MC9S4 was the sample with the lowest aluminium content and the greatest number of acid sites and, therefore, the highest conversion values. Additionally, the esterification of oleic acid with methanol was carried out over a waste material (Amazon flint kaolin) that was thermally treated (at 850 and 950ºC) and activated with sulphuric acid solutions (1 M and 4 M). It was observed that the thermal activation at 950ºC made the flint kaolin more reactive and susceptible to acid leaching. Interestingly, the acid-activated metakaolins presented both Brønsted and Lewis acidic sites. The leached metakaolin treated at 950ºC and activated with 4 M sulphuric acid showed the highest surface area (406 m^2/g), the highest number of acid sites (237.7 mol/g) and offered the maximum esterification activity (98.9%) at 160ºC, an acid:methanol molar ratio of 1:60 and 4 h reaction time. The influence of the molar ratio of the reactants, alcohol chain length, temperature and time was also investigated [61].

The transesterification of triglycerides (triacetin and castor oil) and esterification of ricinoleic acid with methanol was studied over the polymeric catalyst polyaniline-sulphate at 50–60°C. Polymer powder (PANI-S) and three samples of various contents of polymer deposited on a carbon support were studied. All the samples were found to be active solid acid catalysts in both tested reactions. Catalytic performance of polyaniline-sulphate-based catalysts for methanolysis of triacetin (glycerol triacetate), the shortest triglyceride molecule, differed from vegetable oil and castor oil, which consist of long chain triglycerides like ricinoleic acid. In the methanolysis of vegetable oil, the outermost surface of the catalysts was mainly responsible for activity, and much higher activity was exhibited by carbon-supported catalysts with deposited polymer, especially at low polymer content (13.1 wt%). After five successive catalytic runs, activities were found to be ca. 80–95% relative to the activities of fresh catalysts. The esterification of ricinoleic acid with methanol was further carried out over polyaniline-sulphate. It was observed that in the presence of all three catalysts, conversions of 95–98% were achieved. This catalyst thus had the potential to overcome the limitations of sulphated zirconia catalysts and was chosen because of its easy preparation and handling, environmental and thermal stability, and insolubility in most inorganic solvents [62].

VALORIZATION OF WASTE COOKING OIL WITH HIGH AMOUNT OF FFA (OLEIC ACID) INTO BIODIESEL OVER SOLID ACID CATALYSTS

The major obstacle for the commercial production of biodiesel is the high cost of raw material (i.e., refined vegetable oils). To overcome this problem, low-cost feed-stocks, such as waste oils and fats, can be used. However, these feed-stocks contain a high amount of free fatty acids (FFAs), which cannot be used for the production of biodiesel using traditional homogeneous alkali-catalysed transesterification processes. Different solid acids have thus been used as catalysts in the conversion of waste cooking oil with a high amount of free fatty acids into biodiesel.

Resins with Sulfonic Acid Groups

The esterification of FFA in WCO was carried out over acidic ion-exchange resins (Amberlyst-15 (A-15), Amberlyst-35 (A-35), Amberlyst-16 (A-16) and Dowex HCR-W2). The authors studied the effect of temperature (50–60°C) and catalyst loading (1–2 wt%) on FFA conversion. FFA conversion increased with increasing reaction temperature and catalyst amount. The catalytic activities was found to be in the order A-15 > A-35 > A-16 > Dowex HCR-W2, which was related to the size of average pore diameters and magnitude of BET surface area. Commercial strong acidic ion exchange resins were also used for esterification of FFA in WCO [63].

Furthermore, Amberlyst-15 was used as a catalyst in the esterification of palm fatty acid distillate, a by-product from the palm oil refining process. No methanol poisoning was observed for Amberlyst-15, even at a methanol content of 25 wt%. Different organic solvents were also tested, but their effect on the yield in the Amberlyst-15 catalytic system was negligible. The maximum biodiesel yield obtained using Amberlyst-15 was 97% [64].

Biodiesel production was recently observed over a cation exchange resin (D002, 002CR and 732) from rapeseed oil deodoriser distillate with 48.80 wt% FFA. Cation exchange resin D002 showed higher activity than 002CR and 732 resins. Biodiesel yield exceeded 96% under the optimal conditions of a D002 resin catalyst dosage of 18 wt% (based on oil weight), an oil to methanol molar ratio of 1:9, a reaction temperature of 60°C, and a reaction time 4 h. the D002 catalyst was reused for 10 cycles without the yield falling below 88% [65].

To prepare biodiesel from acidified oils generated from waste frying oils, three types of cation-exchange resins (NKC-9, 001x7 and D61) have been used as solid acid catalysts. The results showed that the catalytic activity of NKC-9 was higher than that of 001x7 and D61. The esterification by NKC-9 increased with increasing amount of catalyst, reaction temperature and time and methanol/oil molar ratio. The maximal conversion in an optimised reaction was approximately 90.0%. The NKC-9 resin also exhibited good reusability [66].

The production of biodiesel from high FFA oils (oleic acid) has been carried out over two heterogeneous acid catalysts (Amberlyst-15 and Amberlyst BD20). Amberlyst BD20 showed good catalytic efficiency for high FFA oils. When the FFA content of oils was 50.0 and 99.8 wt%, the activity of Amberlyst-15 gradually decreased upon recycling, whereas the activity of Amberlyst BD20 was unchanged. Furthermore, the activity of Amberlyst 15 was inhibited by the water produced during the esterification process, but the activity of Amberlyst BD20 was not similarly affected. Finally, after reuse the activity of Amberlyst BD20 did not decrease [67]. Recently, Tesser et al. reported the esterification of free fatty acids in high-acidity substrates [68].

Different acid ionic exchange polymeric resins (Amberlyst-15, Amberlyst-16, Amberlyst-131 and Relite CFS) were tested in a batch reactor. Two types of substrates were submitted for esterification with methanol in these tests: a model mixture of soybean oil artificially acidified with oleic acid and a commercial high-acidity mixture of waste fatty acids (oleins). Good catalytic activity was observed. Different kinetic runs were subsequently performed on the selected catalysts (Amberlyst-15 and Relite CFS) and the influence of temperature and catalyst concentration was studied.

To maximise the production of biodiesel from soybean soap stock, the effect of water on the esterification of high-FFA oils was investigated. Oleic acid and high acid oil were esterified with methanol in the presence of Amberlyst-15 as a heterogeneous catalyst or sulphuric acid as a homogeneous catalyst. The yield of fatty acid methyl ester (FAME) was studied at oil to methanol molar ratios of 1:3 and 1:6, and reaction temperatures of 60 and 80°C. The rate of esterification of oleic acid significantly decreased as the initial water content increased to 20% of the oil. The activity of Amberlyst-15 decreased more rapidly than that of sulphuric acid due to the direct poisoning of acid sites by water. FAME content of high acid oil (HAAO) prepared from soap stock rapidly increased for the first 30 min of esterification. When methanol and Amberlyst-15 were removed from the HAAO after 30 min of esterification and fresh methanol and a catalyst were added, the time required to reach 85% FAME content was reduced from 6 h to 1.8 h. It was found that the water resulting from the esterification of oleic acid severely inhibited these esterification reactions [69].

Esterification of acidified oil with methanol was carried out over a sulphonated polyethersulphone (SPES)/polyethersulphone (PES) blend catalytic membrane. The free fatty acid conversion reached 97.6% using the SPES/PES catalytic membrane under the optimal esterification conditions. Meanwhile, a SPES/PES membrane with a 20.3% degree of sulphonation showed good catalytic stability. A pseudo-homogeneous kinetic model was established and the results indicated that the reaction rate constant increased with increasing methanol/acidified oil molar ratio, the loading of the catalytic membrane and reaction temperature [70].

An acid ion-exchange sulfonic resin (Relite CFS) was used for esterification of oleic acid with methanol. The reaction was carried out in a batch reactor at different temperatures and in the presence of a certain amount of triglyceride to simulate an oil with a high free fatty acid content (oleic acid, ~50%), which is of great potential interest as a raw material in biodiesel production. The catalyst was a porous copolymer styrene-DVB resin with sulphonic groups (acidity of 3.6 meq/g). The experimental data were interpreted with a second-order, pseudo-homogeneous kinetic model, and a good agreement between the experimental data and the model was obtained [71].

Russbueldt et al. reported the pre-esterification of free fatty acid (FFA) containing vegetable oils over strong acid ion-exchange resins (EBD 100, EBD 200, and EBD 300) [72]. The materials displayed high catalyst activities and high conversions were obtained even without water removal through an extractive esterification. Catalytic activity loss occurred during recycling durability tests. The presented pre-esterification process facilitated the processing of low quality, high FFA oils to biodiesel via basic transesterification and might help reduce production costs.

Biodiesel production from waste cooking oil (WCO) with methanol was also carried out over MCM-41 with sulphonic acid groups (MCM-41-SO_3H), poly(vinyl alcohol) with sulphonic acid groups (PVA-SO_3H) and polystyrene with sulphonic acid groups (PS-SO_3H) at 60ºC. PVA-SO_3H was observed to have the highest catalytic activity.

To optimise the reaction conditions, different reaction parameters were studied. Increased WCO conversion into fatty acid methyl esters (FAME) was observed with increasing PVA-SO_3H loading. When the transesterification/esterification of WCO was carried out with ethanol over PVA-SO_3H at 60ºC, however, a decrease in biodiesel production was observed. WCO conversion into fatty acid ethyl ester (FAEE) increased when the temperature was increased from 60ºC to 80ºC. When different amounts of free fatty acids (FFA) were added to the reaction mixture, a slight increase in the conversion was observed. Moreover, the PVA-SO_3H catalyst was reused and recycled with negligible loss in activity [73].

Zeolites

The removal of free fatty acid (FFA) in waste frying oil by esterification with methanol was carried out using various zeolite catalysts. ZSM-5 (MFI), mordenite (MOR), faujasite (FAU), beta (BEA) zeolites, and silicalite were employed with different Si/Al molar ratios. Oleic acid was added into the waste frying oil as excess FFA to 10 vol% of the oil. The MFI zeolite improved the removal efficiency of FFA through cracking within the pore structure due to its narrow pore mouth.

Additionally, the catalytic activity for FFA removal decreased with decreasing acid strength of the zeolites. The HMOR zeolites in particular exhibited a similar conversion of FFA with different Si/Al molar ratios, despite their acidity being weaker. The acid strength and pore structure of acidic zeolites affected the catalytic activity of FFA removal [74].

Esterification of oleic acid with ethanol was studied over NaY and VOx over USY zeolites. The authors observed that both zeolites had similar initial reaction rates, but a slight difference in final conversion. The NaY zeolite reached its final conversion in a short time, approximately 10 min, while the VOx over USY zeolite reached its final conversion after 50 min of reaction [75].

H+ ion exchanged ZSM-5 (HMFI) and mordenite (HMOR) zeolites were used as catalysts in the esterification of oleic acid in soybean oil with methanol. Conversion of oleic acid was approximately 80% on the HMFI(25) and HMOR zeolites at 60°C.

In terms of pore structure, the bent and narrow pore of the MFI zeolite appeared to slightly increase the conversion through the pore structure effect. The catalytic activity decreased as the acidity of the HMFI zeolites weakened. In addition, the conversion of oleic acid was significantly improved as the amount of acid sites increased [76].

Heteropolyacids

Biodiesel production from waste frying oil with high FFA content was studied over a heteropolyacid ($H_3PW_{12}O_{40}.6H_2O$, PW12). PW12 was shown to be an efficient catalyst for biodiesel fuel production using a lower catalyst concentration (0.0375:1, weight ratio of catalyst to oil), using a lower methanol/oil ratio (70:1) and a lower temperature (65°C). The catalyst was also easily separated from the production mixture and could be reused a minimum of five times [77].

Heteropolyacids were immobilised on mesoporous Ta_2O_5 by a single step sol–gel co-condensation method and used as catalysts for the simultaneously transesterification of triglycerides and esterification of FFAs in soybean oil. The catalyst obtained showed high dispersion of heteropolyacids and could be reused several times [78].

Production of biodiesel from soybean oil (SBO) containing up to 25 wt% palmitic acid (PA) was carried out on a supported heteropolyacid catalyst (PSA). This solid acid catalyst was demonstrated to catalyse both esterification and transesterification simultaneously. At 200°C, a 1:27 oil/alcohol molar ratio, and 3 wt% catalyst loading, a high quality biodiesel with an ester content of 93.95 mass % was produced from SBO feedstock containing 10% PA in 10 h. This single-step solid acid-catalysed process has potential for industrial-scale production of biodiesel from high FFA feedstocks [79].

Ionic Liquids

Esterification of various FFAs (oleic acid, stearic acid, palmitic acid, and myristic acid) with different alcohols was studied using the Brønsted acidic ionic liquid N-methyl-2-pyrrolidonium methyl sulphonate ([NMP][CH_3SO_3]). Satisfactory yields of 93.6–95.3% were observed under optimal reaction conditions. This [NMP][CH_3SO_3] catalyst was also recyclable. This catalyst could also be used in biodiesel production from less expensive feedstocks, such as waste oils, and in quality improvement of biodiesel using various short-chain alcohols [80].

The transesterification of vegetable oils and the esterification of FFAs with alcohols were studied with cerium (III) trisdodecylsulphate trihydrate ($Ce[OSO_3C_{12}H_{25}]_3.3H_2O$). It was observed that the conversion was greater than 95% in less than 4 h for the transesterification of soybean oil with ethanol and methanol.

In addition, the esterification of oleic acid with ethanol demonstrated that the catalyst was able to produce ethyl esters in the presence of water. Furthermore, the recovered catalyst maintained its activity (~95%) over three cycles [81].

Other Catalysts

Esterification of dark oil to fatty acid methyl esters was carried out over WO_3/ZrO_2, H-zeolite and WO_3/Al_2O_3. The WO_3/ZrO_2 catalyst was the most effective catalyst in the esterification of the tested dark oil. The catalytic reaction parameters were optimised so that 20 wt.% WO_3/ZrO_2 had a FFA conversion of 96% at 150°C, 0.4 g/ml (oil), 1:9 (oil to alcohol) molar ratio and 2 h reaction time.

After four cycles, the catalyst showed no loss in activity [82]. Shi et al. reported the esterification of acidified oil with methanol over an organic–inorganic hybrid membrane prepared from zirconium sulphate ($Zr(SO_4)_2$) and sulphonated poly(vinyl alcohol) (SPVA) [83]. The $Zr(SO_4)_2$ particles were found to be more well dispersed in the SPVA matrix as a result of the stronger interaction between $Zr(SO_4)_2$ and SPVA than with the $Zr(SO_4)_2$/ poly(vinyl alcohol) (PVA) hybrid membrane. Esterification results showed that the FFA conversions to fatty acid esters were 81.2% for the $Zr(SO_4)_2$/PVA catalytic membrane and 94.5% for the $Zr(SO_4)_2$/SPVA catalytic membrane. The $Zr(SO_4)_2$/SPVA catalyst also exhibited better catalytic stability. Waste cooking oils containing 27.8 wt% free fatty acids (FFAs) was converted into biodiesel over carbohydrate-derived catalysts prepared from various carbohydrates, including D-glucose, sucrose, cellulose and starch.

The carbohydrate-derived catalysts exhibited higher catalytic activities for both esterification and transesterification compared to two typical solid acid catalysts (sulphated zirconia and niobic acid). Under the optimised reaction conditions, use of the most effective starch-derived catalyst for biodiesel production from waste cooking oils containing 27.8 wt% FFAs afforded 92% yield of methyl esters after 8 h. This catalyst also showed excellent operational stability [84].

Conclusion

Biodiesel, a renewable source of energy, can be a good solution for global energy demands. However, biodiesel is currently not cost competitive with conventional diesel fuel because of its high raw material and production costs. To improve the economics of biodiesel production, cheap, low quality feed stocks, such as waste cooking oil, can be used instead of refined vegetable oil. The use of this type of feed by-product reduces the overall biodiesel production cost.

However, the presence of high amounts of FFAs complicates the use of homogeneous processes (with basic catalysts) in the transesterification. An alternative for the production of biodiesel from oils with high concentrations of FFA is the esterification of free fatty acids before the transesterification of triglycerides.

Traditionally, esterification reactions have been carried out in a homogeneous phase in the presence of acidic catalysts, such as H_2SO_4, HF, H_3PO_4, HCl and p-toluene sulphonic acid. However, homogenous catalysts have many environmental problems. Their separation from reactants and products is also very difficult. Therefore, homogeneous catalysts have been replaced by heterogeneous ones. As a result, different solid catalysts (e.g., resins, zeolites, heteropolyacids, ionic liquids, and activated carbons) have become popular in esterification and transesterification reactions. Employing these methods, biodiesel production could become a "Green Process".

REFERENCES

[1] V. C. McNeff, C. L. McNeff, B. Yan, D. T. Nowlan, M. Rasmussen, A. E. Gyberg, B. J. Krohn, R. L. Fedie and T. R. Hoye, *Appl. Catal. A: Gen.* 343, 39 (2008).

[2] Y. C. Sharma, B. Singh and J. Korstad, *Biofuels, Bioprod. Bioref.* 5, 69 (2011).

[3] J.A. Melero, J. Iglesias and G. Morales, *Green Chem.* 11, 1285 (2009).

[4] M. G. Kulkarni and A. K. Dalai, *Ind. Eng. Chem. Res.* 45, 2901 (2006).

[5] S. Lestari, Päivi Mäki-Arvela, J. Beltramini, G. Q. M. Lu and D. Y. Murzin, *ChemSusChem* 2, 1109 (2009).

[6] Sivasamy, K. Y. Cheah, P. Fornasiero, F. Kemausuor, S. Zinoviev, and S. Miertus, *ChemSusChem* 2, 278 (2009).

[7] F. R. Ma and M. A. Hanna, *Bioresour. Technol.* 70, 1 (1999).

[8] L. C. Meher, D. V. Sagar and S. N. Naik, *Renewable Sustainable Energy Rev.* 10, 248 (2006).

[9] J. M. Marchetti, V. U. Miguel and A. F. Errazu, *Renewable Sustainable Energy Rev.* 11, 1300 (2007).

[10] S. Zullaikah, C.-C. Lai, S. R. Vali and Y.-H. Ju, *Bioresour. Technol.* 96, 1889 (2005).

[11] V. G. Deshmane, P. R. Cogate and A. B. Pandit, *Ind. Eng. Chem. Res.* 48, 7923 (2009).

[12] L. Canoira, M. Rodríguez-Gamero, E. Querol, R. Alcántara, M. Lapuerta and F. Oliva, *Ind. Eng. Chem. Res.* 47, 7997 (2008).

[13] R. Sarin, R. Kumar, S. K. Puri, A. A. Gupta, R. K. Malhotra and S. Makhija, Can. Patent, CA 2008/02617614 A1, 2008.

[14] M. Di Serio, R. Tesser, E. Santacesaria, F. Dimiccoli, F. Cammarota and M. Nastasi, *J. Mol. Catal A:Chem.* 239, 111 (2005).

[15] S. S. Dash and K.M. Parida, *J. Mol. Catal. A:Chem.* 266, 88 (2007).

[16] B. Chemseddine and R. Audinos, *Catal. Today* 25, 417 (1995).

[17] B. Chemseddine and R. Audinos, *J. Memb. Sci.* 115, 77 (1996).

[18] S. Steinigeweg and J. Gmehling, *Ind. Eng. Chem. Res.* 42, 3612 (2003).

[19] S. Pasias, N. Barakos, C. Alexopoulos and N. Papayannakos, *Chem. Eng. Technol.* 29, 1365 (2006).

[20] C.-H. Su, C.-C. Fu, J. Gomes, I.-M. Chu and W.-T. Wu, *AIChE J.* 54, 327 (2008).

[21] C. V. Grossi, E. O. Jardim, M. H. de Araujo, R. M. Lago and M. J. Silva, *Fuel* 89, 257 (2010).

[22] Petchmala, N. Laosiripojana, B. Jongsomjit, M. Goto, J. Panpranot, O. Mekasuwandumrong, A. Shotipruk, *Fuel* 89, 2387 (2010).

[23] S. M. Son, H. Kimura and K. Kusakabe, *Bioresour. Technol.* 102, 2130 (2011).

[24] C.S. Caetano, L. Guerreiro, I.M. Fonseca, A. M. Ramos, J. Vital, J. E. Castanheiro, *Appl. Catal. A: Gen.* 359, 41 (2009).

[25] J. Y. Park, Z. M. Wang, D. K. Kim and J. S. Lee, *Renew Energ.* 35, 614 (2010).

[26] G. Morales, R. van Grieken, A. Martín and F. Martínez, *Chem. Eng. J.* 161, 388 (2010).

[27] S. Furuta, H. Matsuhashi and K. Arata, *Catal. Commun.* 5, 721 (2004).

[28] K. Kiss, F. Omota, A. C. Dimian and G. Rothenberg, *Top Catal.* 40, 141 (2006).

[29] J. L. Ropero-Vega, A. Aldana-Pérez, R. Gómez and M. E. Niño-Gómez, *Appl. Catal. A: Gen.* 379, 24 (2010).

[30] D. E. López, J. G. Goodwin Jr., D. A. Bruce, S. Furuta, *Appl. Catal. A: Gen.* 339, 76 (2008).

[31] J. C. Juan, J. Zhang and M. A. Yarmo, *Appl. Catal. A: Gen.* 332, 209 (2007).

[32] J. C. Juan, J. Zhang and M. A. Yarmo, *Appl. Catal. A: Gen.* 347, 133 (2008).

[33] J.I. Moreno, R. Jaimes, R. Gómez and M.E. Niño-Gómez, *Catal. Today* 172, 34 (2011).

[34] Sarkara, S. K. Ghoshb, P. Pramanik, *J. Mol. Catal. A: Chem.* 327, 73 (2010).

[35] Jimenez-Morales, J. Santamaria-Gonzalez, P. Maireles-Torres and A. Jimenez-Lopez, *Appl. Catal. A: Gen.* 379, 61 (2010).

[36] A. Costa, P. R. S. Braga, J. L. Macedo, J. A. Dias and S. C. L. Dias, *Microporous Mesoporous Mat.* 147, 142 (2012).

[37] I.V. Kozhevnikov, *Chem. Rev.* 98, 171 (1998).

[38] M. Misono and N. Nojiri, *Appl. Catal.* 64, 1 (1990).

[39] F. Oliveira, L. M. Dezaneti, F. A. C. Garcia, J. L. de Macedo, J. A. Dias, S. C. L. Dias and K. S. P. Alvim, *Appl Catal. A-Gen.* 372, 153 (2010).

[40] K Narasimharao, D. R Brown, A. F. Lee, A. D. Newman, P. F. Siril, S. J. Tavener and K. Wilson, *J. Catal.* 248, 226 (2007).

[41] S. Caetano, I. M. Fonseca, A. M. Ramos, J. Vital and J. E. Castanheiro, *Catal. Commun.* 9, 1996 (2008).

[42] Alsalme, E .F. Kozhevnikova and I. V. Kozhevnikov. *Appl. Catal. A: Gen.* 349, 170 (2008).

[43] L. Xu, X. Yang, X. Yu and G. Y. Maynurkader, *Catal. Commun.* 9, 1607 (2008).

[44] Srilatha, N. Lingaiah, B. L. A. P. Devi, R. B. N. Prasad, S. Venkateswar and P. S. S. Prasad, *Appl. Catal A: Gen.* 365, 28 (2009).

[45] I. Tropecêlo, M. H. Casimiro, I. M. Fonseca, A. M. Ramos, J. Vital and J. E. Castanheiro, *Appl. Catal A: Gen.* 390, 183 (2010).

[46] V. Brahmkhatri and A. Patel, *Appl Catal A: Gen.* 403, 161 (2011).

[47] V. Brahmkhatri and A. Patel, *Ind. Eng. Chem. Res.* 50, 6620 (2011).

[48] Srilatha, Ch. R. Kumar, B. L. A. P. Devi, R. B. N. Prasad, P. S. S. Prasada and N. Lingaiah, *Catal. Sci. Technol.* 1, 662 (2011).

[49] Lotero, Y. Liu, D. E. Lopez, K. Suwannakarn, D. A. Bruce and J. G. Goodwin Jr., *Ind. Eng. Chem. Res.* 44, 5353 (2005).

[50] A. Kiss, A. C. Dimian and G. Rothenberg, *Adv. Synth. Catal.* 348, 75 (2006).
[51] Toda, A. Takagaki, M. Okamura, J.N. Kondo, S. Hayashi, K. Domen and M. Hara, *Nature* 438, 178 (2005).
[52] Hara, *Energy Environ. Sci.* 3, 601 (2010).
[53] Takagaki, M. Toda, M. Okamura, J.N. Kondo, S. Hayashi, K. Domen and M. Hara, *Catal. Today* 116, 157 (2006).
[54] Zong, Z. Duan, W. Lou, T.J. Smith and H. Wu, *Green Chem.* 9, 434 (2007).
[55] L. Peng, A. Philippaerts, X. Ke, J. van Noyen, F. de Clippel, G. van Tendeloo, P. A. Jacobs and B. F. Sels, *Catal. Today* 150, 140 (2010).
[56] L. Geng, Y. Wang, G. Yu and Y. Zhu, *Catal. Commun.* 13, 26 (2011).
[57] Q. Shu, Z. Nawaz, J. Gao, Y. Liao, Q. Zhang, D. Wang and J. Wang, *Bioresource Technol.* 101, 5374 (2010).
[58] M. Ghiaci, B. Aghabarari and A. Gil, *Fuel* 90, 3382 (2011).
[59] Song, Y. Qi, T. Deng, X. Hou and Z. Qin, *Renew Energ.* 35, 625 (2010).
[60] L. A. S. Nascimento, R. S. Angélica, C. E. F. Costa, J. R. Zamian, G. N. R. Filho, *Appl. Clay Sci.* 51, 267 (2011).
[61] L. A. S. Nascimento, L. M. Z. Tito, R. S. Angélica, C. E. F. Costa, J. R. Zamiana and G. N. Rocha Filho, *Appl Catal B: Env.* 101, 495 (2011).
[62] Zieba, A. Drelinkiewicz, E. N. Konyushenko and J. Stejskal, *Appl Catal A-Gen.* 383, 169 (2010).
[63] Özbay, N. Oktar and N. A. Tapan. *Fuel* 87, 1789 (2008).
[64] M. M. R. Talukder, J. C. Wu, S. K. Lau, L. C. Cui, G. Shimin and A. Lim, *Energy Fuels* 23, 1 (2009).
[65] Y. Liu and L. Wang, *Chem. Eng. Process* 48, 1152 (2009).
[66] Y. Feng, B. He, Y. Cao, J. Li, M. Liu, F. Yan and X. Liang. *Bioresource Technol.* 101, 1518 (2010).
[67] Ji-Yeon Park, Deog-Keun Kim and Jin-Suk Lee, *Bioresource Technol.* 101, S62 (2010).
[68] R. Tesser, L. Casale, D. Verde, M. Di Serio and E. Santacesaria, *Chem. Eng. J.* 157, 539 (2010).
[69] Ji-Yeon Park, Zhong-Ming Wang, Deog-Keun Kim and Jin-Suk Lee, *Renew. Energ.* 35, 614 (2010).
[70] W. Shi, B. He and J. Li, *Bioresource Technol.* 102, 5389 (2011).
[71] R. Tesser, M. di Serio, M. Guida, M. Nastasi and E. Santacesaria, *Ind. Eng. Chem. Res.* 44, 7978 (2005).
[72] B.M.E. Russbueldt and W.F. Hoelderich, *Appl. Catal A: Gen.* 362, 47 (2009).
[73] I. Tropecêlo, C. S. Caetano, I. M. Fonseca, A. M. Ramos, J. Vital and J. E. Castanheiro, *Biomass Bioenergy* (submitted).
[74] Kyong-Hwan Chung, Duck-Rye Chang, Byung-Geon Park, *Bioresource Technol.* 99, 7438 (2008).
[75] J. M. Marchetti, A. F. Errazu, *Fuel* 87, 3477 (2008).
[76] Kyong-Hwan Chung, Byung-Geon Park, *J. Ind. Eng. Chem.* 15, 388 (2009).
[77] Cao, Y. Chen, F. Zhai, J. Li, J. Wang, X. Wang, S. Wang, W. Zhu, *Biotechnology and Bioengineering* 101, 93 (2008).

[78] L. Xu, W. Li, J. Hu, X. Yang and Y. Guo, *Appl. Catal B: Environ* 90, 587 (2009).
[79] Baig and T.T. Ng, *Energy Fuel* 24, 4712 (2010).
[80] L. Zhang, M. Xian, Y. He, L. Li, J. Yang, S. Yu and X. Xu., *Bioresource Technol.* 100, 4368 (2009).
[81] F. Ghesti, J. L. Macedo, V. C. I. Parente, J. A. Dias and S. C. L. Dias, *Appl. Catal. A: Gen.* 355, 139 (2010).
[82] Y. M. Park, S. H. Chung, H. J. Eom, J. S. Lee and K. Y. Lee. *Bioresource Technol.* 101, 6589 (2010).
[83] W. Shi, B. He, J. Ding, J. Li, F. Yan and X. Liang, *Bioresource Technol.* 101, 1501 (2010).
[84] W. Y. Lou, M. H. Zong and Z. Q. Duan, *Bioresource Technol.* 99, 8752 (2008).

In: Oleic Acid: Dietary Sources, Functions and Health Benefits ISBN: 978-1-62618-332-2
Editor: Luciano Paulino Silva

Chapter 6

SOURCES OF OLEIC ACID AND THEIR TECHNOLOGICAL AND HEALTH IMPLICATIONS

Rouxléne van der Merwe[1,*] and Arno Hugo[2]
[1]Department of Plant Sciences, University of the Free State, South Africa
[2]Department of Microbial, Biochemical and Food Biotechnology, University of the Free State, South Africa

ABSTRACT

Increased concern about health risks of foods has lead to demands for improvement in the quality control of vegetable oils. Oil quality refers not only to the nutritional properties, but also the functionality of the oil. Optimal oil quality does not exist, as it will depend on the end-use of the oil.

The fatty acid composition is one of the main parameters defining oil nutritional quality. The fatty acid composition of edible oil plays an important role in the shelf life, nutritional, and health properties of the oil. Polyunsaturated fatty acids have been desirable for their role in lowering plasma cholesterol levels. High amounts of polyunsaturated fatty acids are present in edible oils such as soybean, safflower, sunflower and peanut.

However, polyunsaturated oils are susceptible to oxidative deterioration and consequently need additional processing and chemical additives that result in a final product with detrimental implications for human health. Monounsaturated fatty acids and especially oleic acid are naturally stable to oxidation and have good frying stability and shelf-life properties.

In fact, oils rich in oleic acid have positive health implications because of their low saturated fatty acid contents and low *trans*-isomer levels. They offer the consumer the choice of better quality oil with reduced health risks.

* Email address: vandermerwer@ufs.ac.za

Introduction

Vegetable oils are valuable commodities in world trade. Soybean, palm, rapeseed and sunflower seed oil are the dominant commodities and contribute to more than 80% of vegetable oils produced worldwide [1]. They are subjected to specific quality requirements for both food and industrial uses [2]. Due to a continuous demand for new oil types, plant breeders have made great effort to develop those quality features demanded by industry. Traditionally oil quality was measured primarily based on oil content and fatty acid composition. The ideal fatty acid composition depended on the end-use of the oil [3, 4]. However, currently other components of vegetable oils that influence their nutritional and technological properties are being emphasized by oil chemists and nutritionists. The main parameters defining the quality of oil are fatty acid composition, the distribution of fatty acids within the triacylglycerol (TAG) molecule and the total content and composition of natural antioxidants, tocopherols and sterols [5, 6].

Fats and oils consist of saturated, monounsaturated and polyunsaturated fatty acids that occur in different amounts. Animal fats usually contain high amounts of saturated fatty acids, while vegetable oils contain high amounts of monounsaturated and polyunsaturated fatty acids [7]. Polyunsaturated fatty acids including linoleic acid (C18:2) and α-linolenic acid (C18:3) are essential fatty acids that are important for normal growth and development of humans. Essential fatty acids are not normally produced by the human body and should be supplied by the diet [8]. Oleic acid (C18:1) is a monounsaturated fatty acid and though a non-essential fatty acid, it is the most abundant fatty acid in all diets and tissues [9].

Olive oil is a rich source of oleic acid and contains more oleic acid and less linoleic and linolenic acids than other vegetable oils. Other oils that naturally contain high oleic acid levels include pecan oil, peanut oil and palm oil. However, with the demand for improved oil quality for specific food and non-food applications as well as concerns about health risks, much research focuses on the modification of fatty acid composition of various oil crops. Genetic engineering and tissue culture lead to the development and commercialization of vegetable oil products with increased oleic acid contents [2]. These include canola or low-erucic acid (C22:1), mid-oleic sunflower, high oleic sunflower and high oleic safflower oils.

Because of their high unsaturation levels, polyunsaturated oils are unsuitable for certain applications such as deep frying or margarine production [10]. For such applications the oil needs additional processing or chemical additives, which may result in a product that has harmful implications for human health. Additionally, polyunsaturated fatty acids are more susceptible to oxidative rancidity [11]. On the other hand, oils containing high levels of monounsaturated fatty acids such as oleic acid are essentially stable to oxidation and does not require additional processing. High oleic acid oils have good frying stability and shelf-life properties as well as a good flavor [8].

From a nutritional viewpoint, saturated fatty acids are regarded as undesirable for human health and their intake has a detrimental atherogenic effect by raising both total serum and low-density lipoprotein (LDL) cholesterol levels compared to unsaturated fatty acids [12]. By replacing saturated fatty acids with polyunsaturated fatty acids in the diet, the risk for heart disease is reduced. Furthermore, oils rich in oleic acid normally have positive health properties because of their low saturated fatty acid contents and low *trans*-isomer levels [13,

14]. Oils rich in oleic acid are naturally stable and have many applications. They offer the consumer the choice of better quality oil with reduced health risks [15].

LIPID BIOSYNTHESIS

In oilseeds, fatty acids are synthesized at high rates over a short period and used preferentially for TAG synthesis. Vegetable oil synthesis is divided into two distinct, spatially separated reactions, namely fatty acid biosynthesis in the plastids and lipid biosynthesis in the cytosol. More than 20 separate biochemical steps are involved, starting with the first step of fatty acid biosynthesis and ending with TAG production [16].

Fatty acid biosynthesis in oilseed is initiated with acetyl-coenzyme A (CoA) as substrate and is catalyzed by the actions of acetyl-CoA carboxylase and the dissociable fatty acid synthetase (FAS) enzyme complex type III [17]. The sequential actions of FAS III and FAS I produce mainly palmitoyl-acyl carrier protein (palmitoyl-ACP), which is then elongated by two carbon atoms to produce stearoyl-ACP by the FAS II complex. In turn, the stearoyl-ACP is desaturated by the stearoyl-ACP desaturase (SAD) enzyme, which introduces the first double bond in the carbon chain (between carbon atoms 9 and 10) to produce oleoyl-ACP. The acyl-ACPs, palmitoyl-ACP, stearoyl-ACP and oleoyl-ACP are hydrolyzed to free fatty acids that can be exported to the cytosol. These are activated to the corresponding acyl-CoAs, by the action of the acyl-ACP thioesterases [18]. Two types of thioesterases have been identified: Fat A and Fat B [19]. Fat A preferentially acts on long chain fatty acids and has a high specificity for 18:1-ACP. In contrast, Fat B exhibits a higher affinity for the saturated 16:0-ACP and 18:0-ACP [20]. Although these thioesterases influence the final composition of seed oils [21], fatty acid biosynthesis is also influenced by the interaction of thioesterases with key enzymes such as FAS II and SAD [22].

Once acyl-ACPs are hydrolyzed, acyl-CoA synthase incorporates acyl molecules into the pool of acyl-CoA. Oleic acid, once incorporated into phosphatidylcholine, can be desaturated to linoleic acid and then into α-linolenic acid by the actions of oleoyl-phosphatidylcholine desaturase (ODS) and linoleoyl-phosphatidylcholine desaturase respectively. Both enzymes are membrane-bound and located in the endoplasmic reticulum [23]. TAG biosynthesis has been proposed to take place by the Kennedy pathway in some species. Fatty acids are sequentially transferred from CoA to positions *sn*-1 and *sn*-2 of glycerol-3-phosphate, resulting in the formation of phosphatidic acid (PA), a central metabolite. Diacyglycerol (DAG) is released by the defosforilation of PA. A third fatty acid is transferred to position *sn*-3 of DAG to form TAG, catalyzed by diacylglycerol-acyltransferase. Most of the fatty acids produced in the plastid are not immediately available for TAG biosynthesis. Instead, in most oilseeds the major flux of acyl chains enter into phosphatidylcholine (PC) pools where further desaturation and hydroxylation occur. The fatty acids from PC may then become available for TAG synthesis by one of two mechanisms: i) Fatty acids attached to CoA and those on PC may essentially trade places. This acyl exchange probably occurs by the combined reverse and forward reactions of an acyl-CoA:PC acyltransferase [24]. The resulting acyl-CoA may then be used as an acyl donor in TAG synthesis. ii) PC can donate its entire DAG portion for TAG synthesis. In many plants, the synthesis of PC from DAG and cytidine 5'-diphosphate (CDP)-choline appears to be rapidly reversible as catalyzed by the choline-DAG phosphor-transferase. The reversibility of this reaction allows the DAG moiety of PC to become available for TAG synthesis. The flux of acyl chains in the endoplasmic reticulum eventually leads to esterification on all three positions of the glycerol to form TAG [25, 26]. A stereochemical numbering system is used to identify the three positions on the glycerol derivative as *sn*-1, *sn*-2 and *sn*-3 from the top to the bottom with the secondary hydroxyl to the left of the central carbon. Several seed oil analyses have indicated that saturated fatty acids such as palmitic (C16:0) and stearic (C18:0) acids tend to occupy the *sn*-1 position, whereas unsaturated fatty acids linoleic and linolenic are primarily found at the *sn*-2 position. The *sn*-3 position is occupied with variable molecular fatty acid species [17].

Alteration of Oil Composition by Traditional Breeding and Genetic Modification

The functional and nutritional properties of oils are the main drivers for the modification of seed oil [27]. Traditional breeding methods have been combined with biotechnology to broaden the array of possibilities to improve food products. In most cases the fatty acid profile of the oilseed is modified. Genetic varieties that have been developed to create new value-added oils include low saturates for dietary needs, low linoleic for flavor stability, high oleic for health and oxidative stability and high saturates that eliminate the need for hydrogenation [8]. Canola, high oleic safflower, high oleic sunflower, and low saturate soybean oil are the result of extensive conventional breeding programs, based on crossing the available crop germplasm. However, mutation breeding provided a new tool for breeders. Chemical and radiation mutagenesis have been used to obtain suitable germplasm for generating novel traits [2]. Mutations leading to modification of enzymatic activities in the lipid biosynthesis pathway could occur at several levels (desaturase gene deletion or duplication, transcription levels of these genes or accumulation of desaturase gene transcripts) and therefore several mechanisms may explain the oleic acid accumulation in high oleic seeds [28].

The introduction of genetic engineering greatly expanded the ability of the breeder to introduce desired traits, with genes from incompatible crops or other organisms. The metabolic pathways of oils in cells may be altered by means of i) the expression of transgenes, ii) down regulation of endogenous genes or a combination of both: i) The ectopic expression of a transgene involves placing of the coding sequence into a proper genetic context to permit the desired transcriptional regulation. Approaches to obtain transgenic plants include the somatic embryogenesis-based system, which involves either *Agrobacterium* -mediated transformation or particle bombardment [29, 30]. ii) Down regulating endogenous gene expression in plant cells is typically carried out by inducing post-transcriptional gene silencing (PTGS). PTGS involves the introduction of a gene or part of a gene that is homologous to the endogenous gene targeted for gene silencing, into the plant cell. Different metabolism enzymes have been cloned and characterized. Microsomal oleoyl-phosphatidylcholine desaturase or *FAD2* is responsible for the conversion of oleic acid to linoleic acid and increased levels of oleic acid have been obtained in various seed oils by altering the function of this enzyme using PTGS technology [31-34].

Crop genetic engineering holds great promise as a means of developing oilseed crops with unique characteristics. Genetically improved oilseed crops may have improved nutritional and commercial value and hold environmental benefit [2].

Edible Oil Sources Rich in Oleic Acid Natural Oil Sources

Olive Oil

Olive oil is obtained from the mesocarp of the fruits of the olive tree (*Olea europea* L.). Commercial growth of the tree is confined almost entirely to the Mediterranean countries of Italy, Greece, Spain, Turkey and Tunisia [2]. The approximate oil content of the olive fruit is 19.6%. The oil is characterized by a high level of oleic acid with fatty acid levels of 55-83% for oleic acid, 4-21% for linoleic acid and 8-20% for palmitic acid according to the *Codex* standards. The oil additionally contains a range of unsaponifiable constituents that confer high oxidative stability [2]. Olive oil is the primary source of fat in the Mediterranean diet and epidemiological studies showed that Mediterranean countries have a low risk of coronary heart disease (CHD) [35]. Olive oil has been shown to reduce the risk of thrombogenesis. The beneficial properties of olive oil are often attributed to its high monounsaturated fatty acid content and especially oleic acid content [36].

The high content of monounsaturated fatty acid in olive oil reduces total and LDL cholesterol levels in comparison with saturated fatty acids. However, other components such as polyphenols and tocopherols also have a positive impact on cardiovascular health [37].

Palm Oil

The oil palm (*Elaeis guineensis* Jacq.) produces two distinct oils namely palm oil which comes from the fleshy endosperm and palm-kernel oil extracted from the kernels. The oil palm grows in the tropical regions of Asia (mainly Malaysia and Indonesia), Africa and America. Palm fruit bunches contain between 200-2000 individual fruits that deliver 20-24% palm oil and 2-4% palm-kernel oil. Palm oil contains almost equal proportions of saturated and unsaturated fatty acids. Palmitic is the predominant fatty acid (44-48%) and is followed by relatively high oleic acid (37-40%) and low linoleic (10-11%) and stearic acid (4%) contents. Palm oil is used in many applications such as cooking oils, margarines, shortenings, vanaspati, salad dressings, mayonnaise and other specialist fats. The oil is naturally stable to oxidation because of its low contents of linoleic and linolenic acids and because it is semi-solid in nature, there is almost no need for hydrogenation. Palm oil can be fractionated into two parts in order to extend the usefulness of the oil. The solid fraction, palm stearin (30-35%) is less favorable, but can be used as a hard fat in the production of spreads. The liquid fraction, palm olein (65-78%) is of high quality and is highly stable for use as frying oil. Through seed breeding, palm trees are being developed with higher oil yields and increased unsaturated oil composition [2, 38].

Tree Nut Oils

Tree nuts are cultivated for use as oil crops in many parts of the world. They are important sources of energy and essential dietary nutrients and are valued for their sensory, nutritional and health attributes. The oil is also used in the production of some skin moisturizers and cosmetic products [39]. In many cases tree nuts provide rich sources of food lipids, up to 75% lipid on a weight basis. Generally tree nut oils are rich in monounsaturated fatty acids, predominantly oleic acid [40]. However, the tree nut oils contain much lower amounts of polyunsaturated fatty acids (linoleic acid) and small amounts of saturated fatty acids. Clinical trials with tree nuts have reported favorable effects on blood lipids and lipoproteins. The beneficial health effects of tree nuts and nut oils are attributed to their high oleic acid content [2, 39, 41]. However, other bioactive constituents in nuts have additional cholesterol lowering effects [42].

Hazelnut Oil

Hazelnuts (*Corylus avellana* L.) are mainly produced in Turkey which account for almost 75% of the global production but Italy (10%), Spain (2.32%) and North America (less than 5%) also contribute to the world production. Hazelnuts are a rich source of energy with an oil content of 50-67% [43].

The lipid profile of hazelnut oil seems to be similar to olive oil. The major fatty acids found in hazelnut oil are oleic (76-84%), linoleic (6-14%), palmitic (4-5%), stearic (2-3%) and linolenic (less than 2%) acid [2, 43]. Although hazelnuts are generally consumed as a fruit, they have various other applications too. They are mainly used as ingredients in confectionary products and as raw materials for the pastry and chocolate industry. Hazelnuts are also used to add flavor to several sweet and savory products for instance bakery, cereal and desert formulations.

The oil may also serve as a source of cooking oil because of its unique flavor and lipid composition [39]. The high monounsaturation of hazelnut oil has favorable effects on both total and LDL cholesterol levels. When hazelnuts were incorporated in the diet, total and LDL cholesterol was significantly reduced [41].

Almond Oil

Almond, *Prunus dulcis* (Miller) D.A. Webb, syn. *Prunus amygdalus* Batsch is mainly produced in Italy and California but other parts of Europe, Asia and Australia also contribute to almond production. The almond, the fruit of the almond tree (*Prunus delcis* and *Prunus amara*) is regarded as nutritionally valuable and therefore the almond has consistently gained commercial importance.

The almond contains on a weight basis, a high level of oil (50.6%). Oleic acid is the predominant fatty acid in almond oil (68%) and is followed by linoleic (25%), palmitic (4.7%) and small amounts of stearic, palmitoleic (C16:1) and arachidic (C20:0) acid [44].

Almond oil is regarded as health-promoting edible oil. Several studies revealed that consumption of almond may improve blood lipid profiles by lowering total and LDL cholesterol and raising plasma high density-lipoprotein (HDL) cholesterol levels [41, 45].

Pecan Oil

Pecan tree [*Carya illinoinensis* (Wangenh.) K. Koch] is native to the United States, but it is commercially produced in Australia, South Africa and several Middle Eastern and South America countries. Fat is the major component of all varieties of pecan and comprises 65-79% on a weight basis. Pecan is another rich source of monounsaturated oil. Oleic (55-75%) acid is the predominant fatty acid present in pecan oil and is followed by linoleic (19-33%), palmitic (7%), linolenic (2%) and stearic (2%) acid. Studies revealed that pecan oil is stable to oxidation regardless of its high unsaturation level. Due to the high oleic acid content of pecan oil, the oil is less susceptible to oxidation [2, 46]. Pecans contain a number of cardio-protective compounds including plant sterols, vitamin E, folic acid, calcium, magnesium, phosphorus, zinc, vitamin A and several B vitamins. Diets enriched with pecan were found to favorably alter the serum lipid profiles in the human body and as a result, reduce the risk of cardiovascular disease [41].

Genetically Modified Vegetable Oils

The fatty acid composition of various oilseeds has been successfully altered by traditional breeding methods or genetic engineering in order to improve the functional and nutritional properties of the oil [27]. Oils rich in oleic acid are considered high oleic oils and have been developed by traditional breeding methods or by genetic engineering. These include variants of regular sunflower, safflower, soybean, canola and peanut [2]. Fatty acid contents vary slightly between different reports for traditional, high oleic and mid-oleic variants of vegetable oils. This can be attributed to differences in genetic backgrounds and growing conditions of the different oil plants.

High Oleic Sunflower Oil

The seed oil of traditional sunflower (*Helianthus annuus* L.) is characterized by a high concentration of linoleic acid (66-72%), a moderate level of oleic acid (16-20%) and low level of linolenic acid (less than 1%). The saturated fatty acids (SFA), palmitic and stearic acids, account for less than 15% of the fatty acids. Lauric (C12:0), arachidic (C20:0), behenic (C22:0), lignoceric (C24:0) and eicosenoic (C20:1) acids occur in minor percentages [47]. Sunflower cultivars with high oleic acid content were introduced during the 1980s. Chemical mutagenesis [48] and selective breeding were employed in sunflower and raised the oleic acid from 29% to 84% creating high oleic acid oil that is now extensively used in both commercial and domestic frying applications. High oleic sunflower oil contains between 75-85% oleic acid [49], a low concentration of linoleic acid (2-9%) and generally less than 10% SFAs.

High oleic sunflower oil is regarded as a speciality oil and has been found useful in food products such as spray oils, cooking oils and other products that require oil with high oxidative stability [8]. The high level of monounsaturation makes the high oleic oil less susceptible to oxidative degradation than the traditional sunflower oil and therefore the high oleic oil shows potential for applications requiring a high oxidative stability [47].

Nusun Oil

NuSun, a mid-oleic sunflower oil variety, was developed by plant geneticist Jerry F. Miller and biochemist Brady A. Fick in 1995 by traditional breeding methods. The oil contains 43-72% oleic acid and between 18-45% linoleic acid [50]. NuSun is recommended for deep-frying purposes because the oil is resistant to oxidation from extended use [51]. The oil does not require hydrogenation for industrial frying applications. Kiatsrichart et al. [52] compared the pan-frying ability of NuSun to commercial frying oil and discovered that NuSun and canola oil differed little in either oxidative stability or fried food flavor. NuSun is furthermore used in the production of zero-*trans* Frito-Lay snacks [53].

Safflower Oil

Safflower, *Carthamus tinctorius* L., is among the oldest crops known and is a member of the Compositae, family Asteraceae. There are some 25 species in the genus distributed from Spain across North America and West Asia to India. Many species are indigenous to the Mediterranean region. Traditionally safflower oil is linoleic-rich with a high linoleic acid content of 77% and low oleic acid (15%) content. The palmitic acid and stearic content is 5.2% and 1.5% respectively [2]. Researchers at the University of California discovered a natural mutation in the safflower plant that resulted in increased oleic acid levels (74-80%) [8]. However, an oleic-rich variety (74% oleic acid) has been developed and is branded as "saffola" [54].

Traditional safflower oil rich in linoleic acid is utilized in various food products such as mayonnaise, salad dressings and soft margarine. However, for some of these applications, the oil needs to be hydrogenated to a more stable form. Traditional safflower oil is considered relatively unstable due to the high linoleic acid content and low level of natural antioxidants in the oil [2, 8]. However, high oleic safflower oil, with reduced linoleic acid content, has shown a significant increase in oxidative stability over traditional safflower oil. The oxidative stability, measured by the active oxygen method (AOM) stability analysis, has increased by three and a half time. High oleic safflower oil proved to be excellent frying oil and showed acceptable fried food stability [55-57].

Soybean Oil

Soybean oil is obtained from soybeans, *Glycine max* L. (Merrill) and the seed contains approximately 40% protein and 20% oil. The soybean crop has the highest protein content and is the largest producer of vegetable oil in the world. The major soybean production areas include the United States, Brazil, Argentina and China [2]. Traditional soybean oil is of high quality and primarily consists of linoleic acid (53%, *Codex* range 48-59%), followed by oleic acid (23%, range 17-30%), palmitic acid (11%, range 8-14%), linolenic acid (7%, range 4-11%) and stearic acid (4%, range 2-5%). Modified soybean oil with high oleic acid content has been developed. High oleic soybean oil contains more than 80% oleic acid with low levels of saturated fatty acids (9%) and linolenic acid (2%) [2, 32, 58]. By exploiting the variation found in soybean germplasm and with conventional breeding, genotypes with mid-oleic acid phenotypes (30-70% oleic acid) have been developed [59-61], but commercial planting of these genotypes has never materialized due to the yield drag associated with the trait [62-64]. In addition, high oleic soybean lines developed by conventional plant breeding have limited commercial value due to the large influence of the environment on the oleic acid content. With mutation breeding the mutant line M23 was obtained using X-ray irradiation of dry seeds [65]. This line was found to have a two fold increase in oleic acid content (up to 54%) in the seed oil compared to the original variety (Bay) with approximately 23% oleic acid content [59]. Through the use of biotechnology, problems with yield drag and environmental instability of oleic acid content in high and mid-oleic acid mutants have largely been solved. Genetic approaches have been applied to enrich soybean oil with oleic acid [66, 67]. Soybean seed oil with oleic acid contents of more than 80% of the total oil has been generated by the DuPont Company [68]. The high oleic acid trait was obtained by down-regulating the expression of the *FAD2* genes that are responsible for the conversion of oleic acid to linoleic acid [31, 69, 70]. Traditional soybean oil is used in many food applications, but the major products are cooking and salad oils. In order to be used as frying and salad oils, the oil is partially hydrogenated due to its high polyunsaturation level. More extensive hydrogenation is necessary for the production of cooking fats and spreads. For the preparation of salad dressing, soybean oil can be used without hydrogenation [2, 38]. Apart from the application of soybean oil in the food industry the oil has industrial application and is used in biodiesel blends. Soybean derived biodiesel shows, among other positive attributes, enhanced biodegradation, increased flashpoint, reduced toxicity and increased lubricity [58].

The high levels of polyunsaturated fatty acids and especially linolenic acid content make soybean oil relatively unstable to oxidation and shorten its shelf life [2, 71]. However, soybean oil that naturally contains a high level of oleic acid is highly desirable because oxidative stability is increased without any undesirable effect on the oil's nutritional and functional properties [71, 72]. Mid-oleic oil should have improved shelf life and flavor quality when used as a salad oil. Mid-oleic soybean oil can also replace hydrogenated oils in light-duty frying applications. The high oxidative stability is desirable for various food and industrial applications. High oleic soybean oil would have stability at high frying temperatures and would reduce the need for hydrogenation. Consequently *trans* fatty acids are reduced. High oleic soybean oil can also be applied for the manufacturing of biodegradable soy-diesel, lubricants and hydraulic oils [69, 73].

Canola Oil

Canola (*Brassica napus* L.) oil is the name for rapeseed oil with substantially lowered amounts of eruric acid and glucosinolates. Canola oil was developed from the original high eruric acid rapeseed oil by Canadian breeders.

This improvement produced premium quality oil with eruric acid levels below 2%. Canola is mainly produced and developed in Western Europe, China, India and Canada. Canola seeds contain 38-44% oil which is characterized by a relatively high level of monounsaturated fatty acids. It contains approximately 60% oleic acid (*Codex* range 51-70%) and is, among the general vegetable oil sources, second to olive oil in oleic acid content. Canola oil contains a low level of saturated fatty acids (less than 7%) which is about half the level present in olive, soybean and corn oil and an intermediate level of polyunsaturated fatty acids.

The polyunsaturated fatty acid content of canola oil is higher than that of olive and palm oil, but lower than that of sunflower soybean and corn oil. The linoleic acid content ranges between 15-30% and linolenic acid between 5-14% [2].

Since the introduction of standard canola, considerable plant breeding efforts have been made to produce canola oil with modified fatty acid composition. Plant breeders have developed canola oil with reduced linolenic acid content (2%), while canola oil rich in oleic acid (ranging between 60-85%) has been developed through genetic engineering [74, 75]. "Monola" is an oleic-rich rapeseed variety with reduced linolenic acid that was developed in Australia. It contains 68% oleic acid [2, 38].

Canola oil is mostly used as a liquid in the non-hydrated form in the production of salad oils and dressings. The liquid oil component is furthermore used in margarine formulations and baking shortenings. Although canola oil is used as cooking oil and for pan-frying, deep fat frying is not recommended. The polyunsaturated fatty acid content makes it unsuitable for deep fat frying, but lightly hydrogenated canola oil stocks are extensively used as stable frying fats due to their low level of polyunsaturation. Lightly hydrogenated oil often appears in blends with other oils to combine the advantages of canola oil with those of other oils.

Additionally the oil has an exceptionally good stability due to its high level of monounsaturation [38]. By reducing the linolenic acid content the frying performance as well as the storage stability of canola was improved. High oleic acid canola oil showed improved frying stability and produced better quality fried potato chips than traditional canola oil [38, 76]. Canola oil has proven to be effective in lowering plasma total and LDL cholesterol levels when saturated fatty acids was replaced with canola oil in the diets of normolipidemic subjects [2].

Peanut Oil

Arachis hypogaea L. is commonly known as peanut or groundnut because the seed develops underground. Peanut belongs to the same legume family as the soybean, but in composition it is more like a nut than a bean. Peanuts are mostly produced in India, China and the United States. Peanut commonly provides food for humans or livestock and forms a valuable protein component in the diet in the absence of meat. Peanut crushing yields about 4.6 million ton of oil each year. Peanut oil is occasionally known as arachis oil or monkey-nut oil.

The total oil content varies between 44-56% depending on cultivar and environmental production conditions. The traditional oil contains relatively high oleic acid content (35-69%) followed by linoleic acid (12-44%) and palmitic acid (8-14%) [2].

Commercial peanut varieties that incorporate the high oleic acid trait have been developed through conventional breeding. The high oleic acid trait was discovered in experimental lines that contained 80% oleic acid and 2% linoleic acid. The trait may be attributed to a natural mutation that reduces oleoyl-PC desaturase activity.

The fatty acid composition of high oleic acid peanut oil is similar to olive oil with increased oleic acid content (up to 80%), reduced linoleic acid (less than 4%) and reduced palmitic acid (9%) [2, 8]. The oxidative stability of peanut oil is highly correlated with the ratio of oleic to linoleic acid. The traditional oil shows high oxidative stability, however, the oxidative stability of high oleic peanut oil proved to be much better than that of traditional peanut oil [8, 77]. It was also shown that high oleic acid peanuts have a two times longer shelf life than traditional peanuts [78]. Traditional peanut oil is mainly used as a cooking, frying and salad oil as well as for the production of shortening, margarine and mayonnaise.

The use of high oleic acid peanuts as the fat source in a low-fat-high monounsaturated diet showed significant positive changes in blood lipids as well as a reduction of total cholesterol in postmenopausal women [79]. It was shown that high oleic peanut oil has positive health benefits by reducing LDL cholesterol and triglyceride levels in the blood. The oil had no effect on HDL cholesterol levels [2, 8].

Cottonseed Oil

The cotton plant is mainly grown for its fiber with the oil being a by-product that represents about 11-12% of the gross value of the product. Crude cottonseed oil is derived mainly from the seeds of *Gossypium hirsutum* L. (American) and *Gossypium barbadense* L. (Egyptian) varieties of cotton. Cottonseed oil occupies ninth place in world production of vegetable oils, and it is mainly grown in China, India, the United States, the previous USSR, Pakistan, Brazil and Turkey. Cottonseed contains 16% oil by weight. Standard cottonseed oil contains a relatively high level of palmitic acid (27%) along with oleic (18-22%) and linoleic acids (51-55%) and a low content of stearic acid (2%) and linolenic acid (less than 1%) [2, 38]. Modified cottonseed oils with a wide range of fatty acid profile options have been produced. Transgenic cotton lines with increased oleic acid contents from 18% to 47% have been developed by researchers at the University of North Texas [80]. Australian researchers have genetically modified the cotton plant to produce seed oil with up to 77% oleic acid [8].

Cottonseed oil is mainly used in salad and cooking oils, shortenings, margarines and speciality products. Traditional cottonseed oil requires less hydrogenation to attain the same degree of hardness than other linoleic oils. However, high oleic cotton seed oil may replace hydrogenated oils in the food service applications. As a result, high oleic cottonseed oil products have potential health benefits due to reduced *trans* fatty acids and also through lowering of serum LDL-cholesterol levels and consequent reduced incidence of cardiovascular disease (CVD) [38, 81].

Corn Oil

Corn oil is a major vegetable oil obtained from the seeds of corn or maize (*Zea mays* L.). It is mainly produced in the United States, the European Union and Japan. Corn seed contain only 3-5% oil which is obtained predominantly through a wet milling process. The germ is a rich source of oil and presents more than 30% of the total oil in the seed. Traditional corn oil is characterized by high levels of linoleic acid (39-63%) and lower levels of oleic acid (20-42%) and palmitic acid (9-17%) and low levels of linolenic acid [2]. In response to the current demand for highly monounsaturated vegetable oils, efforts have been devoted to developing corn hybrids with corn germ oil that contains high levels of oleic acid. A high oleic acid corn hybrid has been patented [82]. This high oleic corn variety has an increased oleic acid level (65%) and reduced linoleic acid content (22%) [8].

Refined corn oil is mainly used as cooking or salad oil. Although traditional corn oil has a high level of unsaturation, the oil shows good stability which can be attributed to the low level of linolenic acid in the oil. For the production of margarines and spreads, partial hydrogenation is necessary. Corn oil has many non-food uses and it is an ingredient in many cosmetic and skincare products. Other industrial applications of corn oil are in paints, varnishes, soaps, textiles and insecticide formulations [2].

Corn oil proved to have cholesterol-lowering properties and this was shown to be the result of the high content of unsaponfiables and mainly phytosterols [83] in the oil. Corn oil has been reported to have beneficial effects on blood pressure and diabetes [84]. Corn oils with increased oleic acid content resulted in higher oxidative stability that is important for frying oil and spray applications [8].

Technological and Health Implications

Oil Oxidative Stability

Lipid oxidation is a major factor in quality deterioration in edible oils and fatty acid foods since it alters their chemical, sensory and nutritional properties [85]. Autoxidation is a major cause of quality losses in crude and refined oils during storage. The rate of oxidation depends on storage conditions, such as temperature and the presence of light, as well as on the availability of soluble and reactive oxygen in the oil's mass [86]. Oil oxidative stability and deterioration depend on the initial oil composition, concentration of compounds with antioxidant or pro-oxidant characteristics and degree of processing [87, 88]. Oxidation of oils occurs at sites of unsaturation [89] and as a result, the rate of oxidation of fatty compounds depends on the number of double bonds and their position [85]. Autoxidation of unsaturated lipids is a series of free radical reactions, initiated and propagated by free radicals reacting with methylene ($-CH_2-$) groups that are adjacent to double bonds. A free radical is an unpaired electron, indicated as a heavy dot in chemical formulas and is a highly reactive species. Autoxidation can be described in terms of initiation, propagation and termination [7, 87, 90].

Initiation starts when a hydrogen atom departs from the α-methylenic carbon, adjacent to a double bond in a fatty acid (RH) group of the lipid molecule. This reaction may be catalyzed by light, heat or metal ions to form a free radical (R•) (reaction 1).

Initiation: $RH \xrightarrow{\text{initiator}} R\bullet + H\bullet$

The resultant alkyl free radical is highly susceptible to attack by atmospheric oxygen and the dissolved oxygen adds to this site and an unstable peroxide free radical is formed (ROO•) (reaction 2). The peroxide free radical abstracts a hydrogen from another methylene group and reacts with the hydrogen to form a hydroperoxide (ROOH) and a new alkyl free radical (reaction 3).

Propagation: $R\bullet + O_2 \rightarrow ROO\bullet$

$ROO\bullet + RH \rightarrow ROOH + R\bullet$

These free radicals serve as strong catalysts of further oxidation reactions, hence oxidative degradation of oils becomes an autocatalytic process. The chain reaction (or propagation) may be terminated by the formation of non-radical products that result from the combination of two radical species (reactions 4-6). In the final stage of oxidation, the hydroperoxides are readily decomposed into aromatic organic compounds, mainly aldehydes, ketones, alcohols and acids. These compounds cause the rancidity condition that ultimately destroys acceptability and usefulness of oils [7, 90-92].

Termination: $R\bullet + R\bullet \rightarrow RR$

$R\bullet + ROO \rightarrow ROOR$

$ROO\bullet + ROO\bullet \rightarrow ROOR + O_2$

The progress of oxidation can be studied by the quantification (or measurement) of oxidized TAG monomers, dimers and polymers [93]. Martín-Polvillo et al. [94] studied the evolution of oxidation in sunflower oils during long-term storage at room temperature and distinguished two oxidation stages: 1) An induction period which is characterized by slow progress of oxidation and 2) an accelerated (advanced) oxidation stage. During the induction period a significant increase in the monomeric oxidation compounds occur and the oxidized monomers are mainly composed of hydroperoxides during the early oxidation stage [93, 94]. The end of the induction period is defined as the point when a notable shift in the oxidation rate is observed and is clearly characterized by a sharp increase in levels of total oxidation compounds, exhaustion of antioxidants (α-tocopherol) and significant formation of polymerization products. The length of the induction period depends on the degree of unsaturation. The higher the degree of oil unsaturation, the shorter the induction period and the higher the amount of primary oxidation products accumulated at the end of the induction period. During the advanced oxidation stage, secondary products are formed. As a consequence, TAG containing oxygenated functions other than the hydroperoxide (epoxy, keto, hydroxy, and others.) start contributing to the amount of oxygenated TAG monomers. Hydroperoxide functions are therefore not only present in primary oxidation compounds but are also involved in dimeric linkages of polymerization compounds [94].

The first high oleic sunflower seed lines were developed by mutagenesis and selective breeding [48]. Compared to traditional sunflower oil, the high oleic sunflower oil has a better oxidative stability for deep-frying applications and extended shelf life properties. In a previous study Martín-Polvillo et al. [94] investigated the oxidative stability of sunflower oils that differed in their unsaturation degrees during long-term storage at room temperature (25ºC). Induction periods obtained for the three oils were compared and a significant correlation was found between the length of the induction period and the degree of oil unsaturation. High oleic sunflower oil proved to be twice as stable as traditional sunflower oil during long-term storage.

The oxidative stability of seed oil of high (82.6% oleic) and mid-oleic acid (53.8% oleic) sunflower lines were compared with those of traditional sunflower (65.7% linoleic and 22.9% oleic). Oxidative stability was evaluated by measuring peroxide, *p*-anisidine and oxidative stability index values. High oleic acid oil had the best oxidative stability. Since oxidative stability was related to the unsaturation level of oil, the high oleic oil with significantly reduced linoleic acid content had the highest resistance to oxidation [95].

Smith et al. [96] also compared the oxidative stability of high oleic sunflower oil with that of traditional sunflower, soybean, maize and groundnut oils during storage at 55°C. Oxidative stability was evaluated by measuring oxygen depletion, head space volatile compound formation and peroxide value. They found that the high oleic sunflower oil had a higher oxidative stability than regular sunflower and soybean oil, but it was similar to those of maize and groundnut oil.

Soybean oil with a lower content of linolenic acid and higher levels of oleic acid was developed by transgenic modification. This high oleic soybean oil has increased oxidative stability, shelf life and performance in deep-frying applications [81, 97-99]. High oleic acid soybean oil with an oleic acid content of 85% and linoleic acid content of 2% were evaluated to determine the oil's frying stability and flavor quality [100]. Results from this study indicated that high oleic soybean oil produced increased oxidative and frying stability compared to low linolenic soybean oil, however, chips fried in high oleic soybean oil had the lowest flavor quality. Blending high oleic soybean oil with low linolenic soybean oil in a 1:1 ratio showed improved flavor quality as well as fried food stability of chips compared to chips fried in high oleic soybean oil.

By lowering the linolenic acid content in canola, the oxidative stability and flavor quality of this oil has been improved. In addition, by decreasing the linolenic acid content and increasing the oleic acid content in canola oil the frying stability of the oil was increased. However, potato chips fried in canola oil with high oleic acid (78%) and low linoleic acid (19%) levels had an undesirable fried-food flavor. Consequently optimal fatty acid composition is necessary to obtain a balance between frying stability, flavor and fried-food quality [101].

The oxidative and flavor stability of high oleic corn oil was studied where frying stability was compared to regular corn oil and other high stability oils [102]. Results showed that dry-milled high oleic corn oil had good initial flavor quality and was significantly more stable than dry-milled traditional corn oil.

Hydrogenation and Its Implication on the Technological Properties of Oils

Hydrogenation involves the chemical addition of hydrogen to unsaturated fatty acids so that the double bonds become saturated. The reaction is carried out by mixing heated oil and hydrogen gas in the presence of a catalyst (a copper or nickel salt). During hydrogenation the double bond may change the configuration and the natural *cis*-isomers may become largely *trans*-isomers. *Trans* fatty acids have a higher melting point than the *cis*-isomers. Full hydrogenation is carried out until all double bonds are saturated; however, some oils are only partially hydrogenated [8, 103].

Hydrogenation changes the physical characteristics of the oil so that the product has improved functionality. Edible oils are often hydrogenated to improve their oxidative or flavor stability. Flavor stability is necessary to maintain product acceptability for long periods after processing and packaging Vegetable oils such as soybean, rapeseed and cottonseed that are highly unsaturated may have limited use in some food applications due to the low oxidative stability of these oils. Partial hydrogenation is mostly used to increase the melting temperature and as a result the oxidative stability of these oils [2, 8].

Trans fatty acids are formed when vegetable oils, rich in polyunsaturated fatty acids, are hardened by hydrogenation to produce fats that have properties such as firmness and plasticity that are required by food manufacturers and the consumer. Large amounts of *trans* fatty acids are found in margarine, margarine-based products, frying oils and fried products [104, 105]. Since the consumption of *trans* fatty acids by humans has been linked to cardiovascular disease, people became more aware of hydrogenated oils and the consequences of their intake [106]. The formation of *trans* fatty acids during hydrogenation is a health concern as hydrogenated plastic fats are widely used in food products [107]. The introduction of genetic engineering in oilseed crops has made it possible to produce oils that are naturally stable and do not need to be hydrogenated. These oils are consequently free of *trans* fatty acids and are considered healthier.

Utilization of Oleic Acid to Manipulate and Improve the Health Properties of Oils and Food Products

Various studies emphasized the importance of reducing total fat and saturated fatty acid intake since reduced intake of these showed beneficial effects on both total and LDL cholesterol. However, diets that contained reduced fat intake resulted in decreased HDL cholesterol levels and raised plasma TAG concentrations [108-110]. The request to reduce the use of saturated fatty acids such as in palm oil may cause an increase in the consumption of *trans* fatty acids because *trans* fatty acids are regarded as the best alternative to saturated fatty acids for the production of semi-solid and solid fats. However, *trans* fatty acids have shown the same unfavorable effect on the serum lipoprotein profile as saturated fatty acids. *Trans* fatty acids not only raise LDL cholesterol and triglyceride levels, but also lower HDL cholesterol levels [111, 112]. In addition, high intake of *trans* fatty acids also decreases insulin sensitivity and increases the risk of type 2 diabetes [113].

Diets replacing saturated fat with monounsaturated fatty acids have a favorable effect on the ratio of total cholesterol to HDL cholesterol, which is a more accurate indicator for CHD risk than total cholesterol alone [110]. Consumption of monounsaturated fatty acids reduces the LDL levels without adversely affecting the HDL cholesterol fraction and as a result reduces the risk for CHD [108, 109].

Several sources of monounsaturated fatty acids are available and include vegetable and tree nut oils that naturally contain high levels of oleic acid as well as oil sources that have been modified through breeding and genetic engineering to contain high levels of oleic acid. Studies have been conducted to determine the effects of high monounsaturated diets on the risk of CHD and most of these demonstrated that a diet high in monounsaturated fatty acids has favorable effects on the CHD profile [110].

There is increasing evidence that consumption of tree nuts and peanuts may be beneficial in decreasing the risk of CHD [110, 114]. Tree nut intake has favorable effects on blood lipids and lipoproteins, major risk factors for CVD when they replace dietary saturated fatty acids [41]. Most tree nuts are low in saturated fat and high in monounsaturated fatty acids. The reduced disease risk was previously solely ascribed to the high monounsaturated fatty acid content of the nut oil, but in later studies it became evident that the presence of other bioactive trace components (phytosterols, squalene and tocopherols) in the nuts may also contribute to the lowered CHD risk [41, 114, 115].

Animal products have been criticized for being unhealthy due to their high saturated fatty acid content. Enhancement of the fatty acid composition of animal products can make a positive contribution to human health [116]. There is a current interest in manipulating the fatty acid profile of poultry meat and eggs mainly due to the imbalances in human dietary intake of various fatty acids as well as a demand for healthier animal protein sources by the consumer [117]. Through modification of the diet of the animal, the proportion of unsaturated fatty acids in the meat, milk and eggs obtained from livestock can be increased. Lipid is supplied in the diet of the animal through various forage sources, oil-rich cereals, fish oil and recently, tallow [116].

Enrichment of poultry diets with vegetable oils has been shown to have different effects on abdominal fat and the site of fatty acid deposition, depending on the fatty acid composition of the oil. The fatty acid profile of the deposited fat in the broiler carcass reflects the dietary fat source. Broiler chickens that received a basal diet supplemented with olive oil resulted in the highest oleic acid content in the deposited fat. The saturated fat was replaced by oleic acid [118]. A study was conducted to determine the effect of various dietary lipid sources including high oleic sunflower oil, standard sunflower oil, fish oil and tallow on the production performance of male broilers. Results of this study suggested that the inclusion of a high level of unsaturated fatty acids (such as high oleic oil), in contrast to saturated fatty acids (tallow), would result in an improved efficiency in conversion of feed mass into body mass in male broiler chickens [117]. The pork meat market has been subjected to several changes during the past few years in order to produce leaner and healthier meat. Strategies have been followed to modify the lipid fraction of the meat. By increasing the unsaturation level of the pork meat lipid fraction, a positive impact on human health is expected. The effect of supplementing the diet with unsaturated fatty acids on the fatty acid profile of the pork meat has been studied [119, 120]. By supplementing high oleic sunflower oil in the diet, the fatty acid composition of the pork meat was changed so that it contained higher levels of oleic acid. Enrichment of swine feed with high oleic sunflower oil is an appropriate strategy to produce pork meat with a suitable oxidative stability [120]. Milk and dairy products are a good source of calcium and are generally perceived as healthy foods. However, milk fat contains approximately 70% saturated fatty acids (mainly myristic (C14:0) and palmitic acids) and these may raise total and LDL cholesterol levels, increasing the risk of CVD. Consequently the consumption of milk fat should be limited [121]. A number of companies use food technology to substitute milk fat with unsaturated fatty acids to produce healthier milk. By substituting saturated fat in dairy products with oleic acid and/or polyunsaturated fatty acids, the risk of CVD is reduced [122, 123]. Recent studies indicated that intake of skimmed milk with omega-3 polyunsaturated fatty acids, oleic acid, folic acid and vitamin E has favorable effects on cardiovascular health. This milk proved to decrease triacylglycerol, total cholesterol and LDL cholesterol levels in the blood [124, 125].

The inclusion of oilseed in the diet of dairy cows has shown to have an influence on the fat composition of the milk. By supplementing the diet of cows with full fat rapeseed or full fat soybean on pasture reduced the solid fat content of the milk at low temperatures. Supplementation with oilseeds altered the fatty acid composition of the milk, resulting in increased oleic acid levels. This change in fatty acid composition resulted in a softer milk fat from which monounsaturated fatty acid butter, that is spreadable at refrigeration temperatures, can be produced [126]. It has been reported that high polyunsaturated fatty acid content in milk increases the risk of oxidation and off flavor development [127]. However, increases in the monounsaturated fatty acid content of milk fats have no adverse effect on the oxidative stability of butter oil [128].

Conclusion

During the last three decades research has been carried out to improve oil quality in many oil crops. Improved oil quality is necessary for specific nutritional requirements as well as for technological and industrial applications. The fatty acid composition of edible oil plays an important role in its nutritional and health properties as well as its shelf life. Several research and breeding efforts concentrated on the modification of the fatty acid composition. A main focus was increasing the oleic acid content to meet nutritional and technological demands. Diets rich in unsaturated fatty acids have shown favorable effects on cholesterol profiles and CVD risk compared to diets rich in saturated fatty acids. However, polyunsaturated oils are unstable to oxidation and need chemical modification (hydrogenation) in order to be used in many applications such as frying. Whilst hydrogenation stabilizes the oil against oxidation, it lowers the oil's nutritional value and induces health risks. Monounsaturated fatty acid (oleic acid) is more stable in frying and cooking applications and has longer shelf life than polyunsaturated oils. Since high oleic oils are naturally stable they do not need to be hydrogenated and are therefore *trans*-fat free. In addition, high oleic oils proved to lower LDL cholesterol without adversely affecting HDL cholesterol. The inclusion of oilseeds in the diets of poultry, swine and cows has shown an influence on the fat composition of milk and dairy products as well as meat. In return, the fatty acid composition has an influence on the oxidative stability of these products. By supplementing high oleic oils in the livestock diet, the fatty acid composition of the meat milk and dairy products was changed so that it contained higher levels of oleic acid and as a result better stability to oxidation. High oleic oils offer good nutritional and technological quality as well health benefits. By establishing the nutritional and functional quality of all oleic acid sources their use in many food and industrial applications will be increased.

REFERENCES

[1] S. Mielke, The revised oil world 2020- Supply, demand and prices, ISTA Mielke, GmbH, Hamburg, Germany (2002).
[2] F. Shahidi, Bailey's Industrial Oil and Fat Products, John Wiley and Sons, Inc. (2005).
[3] P. F. Knowles, *Econ. Bot.* 37, 423 (1983).
[4] D. Rondanini, R. Savin and A. J. Hall, *Field Crop. Res.*, 83, 79 (2003).
[5] J. M. Fernández-Martínez, L. Velasco and B. Pérez-Vich, "Progress in the genetic modification of sunflower oil quality", Proceedings of the 16th International Sunflower Conference, 29, Aug.-2 Sep. 2004 (Fargo, ND, USA) pp1-14.
[6] J. M. Fernández-Martínez, B. Pérez-Vich, L. Velasco and J. Domínguez, *Helia,* 30, 75 (2007).
[7] C. E. Stauffer, Fats and oils, Eagan Press Handbook Series, American Association of Cereal Chemists, Inc., USA (1996).
[8] R. D. O'Brien, Fats and Oils: Formulating and processing for applications, CRC Press, LLC, Florida (2004).
[9] F. Visioli and C. Galli, *Nutr. Rev.* 56, 142 (1998).
[10] B. S. Ghotra, S. D. Dyal and S. S. Narine, *Food Res. Int.,* 35, 1015 (2002).
[11] J. C. Allen and R. J. Hamilton, Rancidity in foods, Aspen Publishers, Inc. (1999).
[12] M. B. Katan, P. L. Zock and R. P. Mensink, *Am. J. Clin. Nutr.,* 61, 1368S (1995).
[13] R. P. Mensink and M. B. Katan, *N. Eng. J. Med.* ,321, 436 (1989).
[14] P. M. Kris-Etherton and S. Yu, *Am. J. Clin. Nutr.,* 65, 1628S (1997).
[15] R. Yodice, *Eur. J. Lipid Sci. Technol.,* 92, 121 (1990).
[16] T. M. Kuo and H. W. Gardner, Lipid Biotechnology, Marcel Dekker Inc., USA (2002).
[17] J. L. Harwood, *Biochimica et Biophysica Acta.,* 1301, 7 (1996).
[18] H. Rolletschek, L. Borisjuk, A. Sánchez-García, C. Gotor, L. C. Romero, J. M. Martínez-Rivas and M. Mancha, *J. Exp. Bot.,* 58, 3171 (2007).
[19] E. Martínez-Force, S. Cantisán, M. J. Serrano-Vega and R. Garcés, *Planta,* 211, 673 (2000).
[20] R.Pleite, E. Martínez-Force and R. Garcés, *J. Plant Physiol.,* 163, 885 (2006).
[21] H. M. Davies, *Phytochemistry,* 33, 1353 (1993).
[22] E. Martínez-Force and R. Garcés, *Plant Physiol. Biochem.,* 40, 383 (2002).
[23] D. A. Gray and R. G. O. Kekwick, *Plant Sci.,* 119, 11 (1996).
[24] P. K. Stumpf and E. E. Conn, The biochemistry of plants, Academic Press Orlando, F.L. (1987).
[25] J. Ohlrogge and J. Browse, *Plant Cell,* 7, 957 (1995).
[26] J. J. Thelen and J. B. Ohlrogge, *Metab. Eng.,* 4, 12 (2002).
[27] D. D. Bills and S. D. Kung, Biotechnology and nutrition: Proceedings of the third international symposium, Boston, M.A. Butterworth-Heinemann, USA (1992).
[28] S. Lacombe and A. Bervillé, "Analysis of desaturase transcript accumulation in normal and in high oleic oil sunflower development seeds", Proceedings of the15th International Sunflower Conference, 12-15, Jun. 2000 (Toulouse, France) ppA1-A6.
[29] J. J. Finer and M. D. McMullen, *In Vitro Cell Dev. B,* 27, 175 (1991).
[30] H. K. Liu, C. Yang and Z. M. Wei, *Planta,* 219, 1042 (2004).

[31] T. Buhr, S. Sato, F. Ebrahim, A. Xing, Y. Zhou, M. Mathiesen, B. Schweiger, A. J. Kinney, P. Staswick and T. Clemente, *Plant J.,* 30, 155 (2002).
[32] A. J. Kinney and T. E. Clemente, *Fuel Process. Tehnol.*, 86, 1137 (2005).
[33] S. Lacombe, I. Souyris and A. Bervillé, *Mol. Genet. Genomics,* 281, 43 (2009).
[34] W. Chen, K. Song, Y. Cai, W. Li, B. Liu and L. Liu, *Plant Mol. Biol. Rep.,* 29, 866 (2011).
[35] A. Keys, *Am. J. Clin. Nutr.,* 61, 1321S (1995).
[36] H. M. Roche, A. Zampelas, J. M. E. Knapper, D. Webb, C. Brooks, K. G. Jackson, J. W. Wright, B. J. Gould, A. Kafatos, M. J. Gibney and C. M. Williams, *Am. J. Clin. Nutr.,* 68, 552 (1998).
[37] A. H. Stark and Z. M. Madar, *Nutr. Rev.,* 60, 170 (2002).
[38] F. D. Gunstone, Vegetable oils in food technology: Composition, properties and uses, Blackwell Publishing Ltd. CRC Press LLC, USA, Canada (2002).
[39] H. Miraliakbari and F. Shahidi, *J. Food Lipids,* 15, 81 (2008).
[40] M. Venkatachalam and S. K. Sathe, *J. Agricult. Food Chem.,* 54, 4705 (2006).
[41] A. E. Griel and P. M. Kris-Etherton, *Brit. J. Nutr.,* 96, S68 (2006).
[42] P. M. Kris-Etherton, S. Yu-Poth, J. Sabaté, H. E. Ratcliffe, G. Zhao and T. D. Etherton, *Am. J. Clin. Nutr.,* 70, 504S (1999).
[43] J. C. Bada, M. León-Camacho, M. Prieto and L. Alonso, *Eur. J. Lipid Sci. Tech.,* 106, 294 (2004).
[44] Z. Shi, F. Qizhong, C. Bangjie and X. Shuying, *Chinese Journal of Chromatography,* 17, 506 (1999).
[45] D. A. Hydson, B. O. Schneeman and P. A. Davis, *J. Nutr.,* 132, 703 (2002).
[46] J. F. Toro-Vazquez, M. A. Charó-Alonso and F. Pérez-Briceño, *J. Am. Oil Chem. Soc.,* 76, 957 (1999).
[47] G. J. Seiler and M. E. Brothers, *Econ. Bot.,* 53, 273 (1999).
[48] K. I. Soldatov, "Chemical mutagenesis in sunflower breeding", Proceedings of the 7th International Sunflower Conference, 27, Jun.-3, Jul. 1976 (Krasnodar, USSR) pp352-357.
[49] A. A. Schneiter, Sunflower technology and production, Agronomy Monograph no 35. American Society of Agronomy Inc., Crop Sci. Society of America Inc., Soil Science Society of America Inc., Wisconsin, USA (1997).
[50] R. F. Wilson, Proceedings of the World Conference on Oilseed Processing Utilization, American Oil Chemists' Society Press, Champaign, Illinois (2001).
[51] K. Warner, B. Vick, L. Kleingartner, R. Isaak and K. Doroff. Reprint, 2010 at http://www.sunflowernsa.com/uploads/Warner.pdf.
[52] S. Kiatsrichart, M. S. Brewer, K. R. Cadwallader and W. E. Artz, *J. Am. Oil Chem. Soc.,* 80, 479 (2003).
[53] M. B. McClellan. Reprint, 2011 at http://www.sunflowernsa.com/uploads/resources/40/frito-brochure-october-2005.pdf.
[54] J. R. Smith, Safflower, AOCS Press, Champaign, Illinois (1996).
[55] G. Fuller, G. O. Kobler and T. H. Applewhite, *J. Am. Oil Chem. Soc.,* 43, 477 (1966).
[56] G. Fuller, D. G. Guadagni, M. L. Weaver, G. Notter and R. J. Horvat, *J. Food Sci.,* 36, 43 (1971).
[57] G. R. List, T. L. Mounts, F. Orthoefer and W. E. Neff, *J. Am. Oil Chem. Soc.,* 73, 729 (1996).

[58] K. Warner and W. Fehr, *J. Am. Oil Chem. Soc.,* 85, 945 (2008).
[59] Y. Takagi and S. M. Rahman, *Theor. Appl. Genet.,* 92, 179 (1996).
[60] S. M. Rahman, T. Kinoshita, T. Anai and Y. Takagi, *Crop. Sci.,* 41, 26 (2001).
[61] J. L. Alt, W. R. Fehr, G. A. Welke and J. G. Shannon, *Crop. Sci.,* 45, 2005 (2005).
[62] W. R. Fehr, *Crop. Sci.,* 47, S72 (2007).
[63] E. Bachlava, J. W. Burton, C. Brownie, S. Wang, J. Auclair and A. J. Cardinal, *Crop. Sci.,* 48, 1764 (2008).
[64] C. W. Scherder and W. R. Fehr, *Crop. Sci.,* 48, 1755 (2008).
[65] S. M. Rahman, Y. Takagi, K. Kutoba, K. Miyamoto and T. Kawakita, *Biosci. Biotechnol. Biochem.,* 58, 1070 (1994).
[66] A. J. Kinney, *J. Food Lipids,* 3, 273 (1996).
[67] J. K. Setlow, Genetic engineering, New York: Plenum Press (1997).
[68] S. Rollerand and S. Harlander, Genetic modification in the food industry, Blackie Academic and Professional, London (1998).
[69] E. B. Cahoon, *Ag. Bio. Forum,* 6, 11 (2003).
[70] T. E. Clemente and E. B. Cahoon, *Plant Physiol.* 151, 1030 (2009).
[71] L. Yan and P. S. Kerr, *Nutr. Rev.,* 60, 135 (2002).
[72] F. R. Wolf and S. Knowlton, "Edible applications in high oleic soybean oil", The 90th American Oil Chemists' Society Annual Meeting and Expo, 9-12, May 1999 (Orlando, Florida) ppS24.
[73] J. D. Lee, K. D. Bilyeu and J. G. Shannon, *J. Crop Sci. Biotech.,* 10, 210 (2007).
[74] R. Scarth, P. B. E. McVetty, S. R. Rimmer and B. R. Stefansson, *Can. J. Plant Sci.,* 68, 509 (1988).
[75] R. Wong, J. D. Patel, I. Grant, J. Parker, D. Charne, M. Elhalwagy and E. Sys, "The development of high oleic canola", Proceedings of the 8th International Rapeseed Congress, 9-11, Jul. 1991 (Saskatoon, Canada) pp53.
[76] I. Petukov, L. Malcolmson, R. Przybylski and L. Armstrong, *J. Am. Oil Chem. Soc.,* 76, 627 (1999).
[77] S. F. O'Keefe, V. A. Wiley and D. A. Knauft, *J. Am. Oil Chem. Soc.,* 70, 489 (1993).
[78] J. C. Braddock, C. A. Sims and S. F. O'Keefe, *J. Food Sci.,* 60, 489 (2006).
[79] D. J. O'Byrne, D. A. Knauft and R. B. Shireman, *Lipids* 32, 687 (1997).
[80] K. D. Chapman, S. Austin-Brown, S. A. Sparace, A. J. Kinney, K. G. Ripp, I. Pirtle and R. M. Pirtle, *J. Am. Oil Chem. Soc.,* 78, 941 (2001).
[81] Q. Liu, S. Singh and A. Green, *J. Am. Coll. Nutr.,* 21, 205S (2002).
[82] K. J. Leto and J. F. Ulrich, United States Patent No. 6,248,939, June 19, 2001.
[83] R. E. Ostlund, S. B. Racette, A. Okeke and W. F. Stenson, *Am. J. Clin. Nutr.,* 75, 1000 (2002).
[84] W. J. Wechter, United States Patent No. 6,242,479, June 5, 2001.
[85] E. N. Frankel, Lipid oxidation, The Oily Press, Dundee, Scotland (1998).
[86] G. Márquez-Ruiz, M. Martín-Polvillo and M. C. Dobarganes, *Lipids,* 338, 233 (2003).
[87] G. H. Crapiste, M. I. V. Brevedan and A. A. Carelli, *J. Am. Oil Chem. Soc.,* 76, 1437 (1999).
[88] A. Kanavouras, A. Cert and R. J. Hernandez, *Food Sci. Tehnol. Int.,* 11, 183 (2005).
[89] T. P. Labuza and L. R. (Jr.) Dugan, Kinetics *CRC Crit. Rev. Food. Technol.,* 2, 355 (1971).
[90] E. Choe and D. B. Min, *Compr. Rev. Food Sci. F,* 5, 169 (2006).

[91] E. R. Sherwin, *J. Am. Oil Chem. Soc.*, 55, 809 (1978).
[92] F. Shahidi and U. N. Wanasundara, *Food Sci. Tehnol. Int.*, 2, 73 (1996).
[93] G. Márquez-Ruiz, M. Martín-Polvillo and M. C. Dobarganes, *Grasas Aceites,* 47, 48 (1996).
[94] M. Martín-Polvillo, G. Márquez-Ruiz and M. C. Dobarganes, *J. Am. Oil Chem. Soc.,* 81, 577 (2004).
[95] R. Van der Merwe, Genotype by environment interaction for oil quality in high oleic acid sunflower lines, P.hD. Thesis, University of the Free State, Bloemfontein, South Africa (2010).
[96] S. A. Smith, R. E. King and D. B. Min, *Food Chem.,* 102, 1208 (2007).
[97] E. G. Hammond and W. R. Fehr, *J. Am. Oil Chem. Soc.,* 11, 1713 (1984).
[98] T. L. Mount, K. Warner, G. R. List, W. E. Neff and R. F. Wilson, *J. Am. Oil Chem. Soc.,* 71, 495 (1994).
[99] K. Liu, *INFORM* 10, 868 (1999).
[100] K. Warner and M. Gupta, *J. Food Sci.,* 70, S395 (2005).
[101] K. Warner, P. Orr and M. Glynn, *J. Am. Oil Chem. Soc.,* 74, 347 (1997).
[102] K. Warner and S. Knowlton, *J. Am. Oil Chem. Soc.,* 74, 1317 (1997).
[103] C. Alais and G. Linden, Food biochemistry, Ellis Horwood Ltd., England (1991).
[104] L. H. Wiedermann, *J. Am. Oil Chem. Soc.,* 55, 823 (1978).
[105] K. H. Wagner, E. Auer and I. Elmadfa, *Eur. Food Res. Technol.,* 210, 237 (2000).
[106] D. Mozaffarian, M. B. Katan, A. Ascherio, M. J. Stampfer and W. C. Willett, *N. Engl. J. Med.,* 354, 1601 (2006).
[107] F. Shahidi, P. K. J. P. D. Wanasundara and U. N. Wanasundara, *J. Food Lipids,* 4, 199 (1997).
[108] F. H. Mattson and S. M. Grundy, *J. Lipid Res.,* 26, 194 (1985).
[109] R. P. Mensink and M. B. Katan, *N. Engl. J. Med.,* 321, 436 (1989).
[110] P. M. Kris-Etherton, T. A. Person, Y. Wan, R. L. Hargrove, K. Moriarty, V. Fishell and T. D. Etherton, *Am. J. Clin. Nutr.,* 70, 1009 (1999).
[111] R. P. Mensink and M. B. Katan, *N. Engl. J. Med.,* 323, 439 (1990).
[112] J. T. Judd, B. A. Clevidence, R. A. Muesing, J. Wittes, M. E. Sunkin and J. J. Podczasy, *Am. J. Clin. Nutr.,* 59, 861 (1994).
[113] J. Salmerón, F. B. Hu, J. E. Manson, M. J. Stampfer, G. A. Colditz, E. B. Rinum, and W. C. Willett, *Am. J. Clin. Nutr.,* 73, 1019 (2001).
[114] L. S. Maguire, S. M. O'Sullivan, K. Galvin, T. P. O'Connor and N. M. O'Brien, *Int. J. Food Sci Nutr.,* 55, 171 (2004).
[115] P. M. Kris-Etherton, G. Zhao, A. E. Binkoski, S. M. Coval and T. D. Etherton, *Nutr. Rev.,* 59, 103 (2001).
[116] V. B. Woods and A. M. Fearon, *Livest. Sci.,* 126, 1 (2009).
[117] F. H. De Witt, S. P. Els, H. J. Van der Merwe, A. Hugo and M. D. Fair, *S. Afr. J. Anim. Sci.,* 39, 45 (2009).
[118] N. Crespo and E. Esteve-García, *Poult. Sci.,* 81, 1533 (2002).
[119] J. Mitchaothai, C. Yuangklang, S. Wittayakun, K. Vasupen, S. Wongsutthavas, P. Srenanul, R. Hovenier, H. Everts and A. C. Beynen, *Meat Sci.,* 76, 95 (2007).
[120] V. Cardenia, M. T. Rodriguez-Estrada, F. Cumella, L. Sardi, G. D. Casa and G. Lercker, *Meat Sci.,* 88, 271 (2011).

[121] A. H. Lichtenstein, L. J. Appel, M. Brands, M. Carnethon, S. Daniels, H. A. Franch, B. Franklin, P. Kris-Etherton, W. S. Harris, B. Howard, N. Karanja, M. Lefevre, L. Rudel, F. Sacks, L. Van Horn, M. Winston and J. Wylie-Rosett, *Circulation,* 114, 82 (2006).

[122] J. Fonollá, E. López-Huertas, F. J. Machado, D. Molina, I. Álvarez, E. Mármol, M. Navas, E. Palacín, M. J. García-Valls, B. Remón, J. J. Boza and J. L. Marti, *Nutrition,* 25, 408 (2009).

[123] E. Lopez-Huertas, *Pharmacol. Res.* 61, 200 (2010).

[124] L. Baró, J. Fonollá, J. L. Peña, A. Martínez-Férez, A. Lucena, J. Jiménez, J. J. Boza and E. López-Huertas, *Clin. Nutr.,* 22, 175 (2003).

[125] P. Benito, J. Caballero, J. Moreno, C. Gutiérrez-Alcántara, C. Muñoz, G. Rojo, S. Garcia and F. C. Soriguer, *Clin. Nutr.,* 25, 581 (2006).

[126] J. J. Murphy, J. F. Connolly and G. P. McNeill, *Livest. Prod. Sci.,* 44, 13 (1995).

[127] M. Focant, E. Mignolet, M. Marique, F. Clabots, T. Breyne, D. Dalemans and Y. Larondelle, *J. Dairy Sci.,* 81, 1095 (1998).

[128] A. M. Fearon, C. S. Mayne and C. T. Charlton, *J. Sci. Food Agric.,* 76, 546 (1998).

In: Oleic Acid: Dietary Sources, Functions and Health Benefits ISBN: 978-1-62618-332-2
Editor: Luciano Paulino Silva

Chapter 7

OLEIC ACID IN MILK OF DIFFERENT MAMMALIAN SPECIES

Paolo Polidori[1]* and Silvia Vincenzetti[2]
[1]School of Pharmacy, University of Camerino, Matelica (MC), Italy
[2]School of Veterinary Medical Sciences, University of Camerino, Via Circonvallazione, Matelica (MC), Italy

ABSTRACT

Milk, being a biological fluid, varies considerably from different mammal species in several of its constituents. Fat content of milk is the nutrient showing the most variation in its percentage. In fact, milk fat is one of the most complex of all common fats, composed of about 98% triglycerides; other milk lipids (as a percentage of the total milk fat) include diacylglycerides (0.25-0.48%), monoacylglycerides (0.02-0.04%), phospholipids (0.6-1.0%), cholesterol (0.2-0.4%), glycolipids (0.006%) and free fatty acids (0.1-0.4%). Milk fat contains many fatty acids, mainly saturated but also monounsaturated and polyunsaturated. Several factors, many of which are interactive, are known to cause variation in milk fatty acid composition. These include stage of lactation, seasonal variations, feeding and animal genetics. Milk's fatty acid composition has been frequently investigated because of its implications for human health; due to the relationship between high-fat diets and heart diseases, consumer interest in milk fat content and also in fatty acid composition of foods has grown in the recent years. Oleic acid (C18:1cis9), a monounsaturated fatty acid, has been deeply studied because of its possible beneficial effects on human health, normally associated with a reduction of serum cholesterol and plasma cholesterol levels. Scientific evidence and nutritional guidelines recommend a reduction in total fat intake, and particularly of saturated fatty acids, which are associated with an increased risk of obesity, hypercholesterolemia and some cancers. In mammalian species, milk fat has a double origin: circulating fatty acids are transported in lipoproteins, other fatty acids are synthesized in the udder. The fatty acid composition of milk fat is of interest because it affects both the texture and the flavor of dairy products. Milk fat in ruminant species is characterized by a higher content of saturated fatty acids, whereas fatty acid composition in mare's milk and donkey's milk

* E-mail address: paolo.polidori@unicam.it.

shows interesting percentages of essential fatty acids. Oleic acid is the most abundant MUFA in all the different milk samples considered in this chapter, and its importance to prevent cardiovascular diseases and some inflammatory disorders is confirmed by different clinical studies.

INTRODUCTION

All the foods of animal origin used by human beings over the centuries hold a more-or-less important lipid fraction. Only a few decades ago lipid content was considered as a good and hence desirable food nutrient. The preference for fatty foods was due to the different life style of our ancestors: work and transportation were not supplied by machines but by muscular activity; therefore a diet with a high energy concentration was necessary. Currently, the situation has completely changed: the longer average life expectancy makes our vascular system more susceptible to atherogenesis and thrombogenesis.

There is a strong relation between the consumption of animal products, their composition and the incidence of chronic diseases in humans, although it has been difficult to establish causal relations among these factors [1]. There is epidemiological evidence positively correlating the incidence of cardiovascular diseases and lipid consumption, in particular cholesterol and fatty acids. Saturated fatty acids represent 58 to 82% of total milk lipids, and they have been associated with hypercholesterolemia and increased levels of low-density lipoproteins (LDL), particularly in the case of palmitic, myristic and lauric acids [2]. On the other hand, dietary mono (MUFA) and polyunsaturated acids (PUFA) would reduce those risks, increasing high-density lipoproteins, HDL [3].

Dietary fat has been implicated with coronary heart diseases, atherosclerosis, diabetes and other health problems. Various epidemiological studies including several countries showed that the Mediterranean region had a significantly reduced all-cause death rate [4]. This reduction was mainly due to the decreased incidence of degenerative disease as well as cancer and coronary heart disease (CHD). This protective effect was assigned to the Mediterranean diet, rich in healthful fiber, fish, fruits, nuts, seeds, vegetables and olive products. A large randomized clinical trial demonstrated that a modified diet according to the Mediterranean standard reduced total deaths by 56% and decreased cancer risk by 61% at a 4-year follow up. Other diets, similarly rich in fruits and vegetables but lacking the olive component, were not equally protective. In the past, the healthy role of olives was proposed, and this hypothesis was supported by the observation that Mediterranean populations consume 20 times more olive oil than Americans; consequently their cancer risk is at least halved. On the other hand, if compared to diets of other Countries, the Mediterranean diet shows a relatively high fat content; the type of fat appears to be more important than the total amount consumed [5].

The need to use natural food products that safeguard biodiversity and that have an application in clinical and nutritional fields is strongly felt by many consumers, and milk can be considered the most common food for a large amount of the human population in the entire world. For this reason, in this chapter, the fatty acid composition will be described, with particular regard to oleic acid content, in milk obtained from different mammalian species, two ruminant species such as dairy cows and sheep, and two monogastric species such as mares and donkeys. Oleic acid is always the most represented fatty acid in all the different kinds of milk described in this chapter: the importance and the biological role of oleic acid will be considered for milk obtained from dairy cows, ewe, mare and donkey.

Milk Fat

Fat is an important component of milk for a number of reasons. Nutritionally, fat is the major energy component of milk, and it accounts for many of the physical properties, manufacturing characteristics and sensory attributes of milk and dairy products [6]. The composition of milk fat is influenced by animal and environmental factors such as breed, diet, stage of lactation, season of year, local temperature and body condition. Approximately 95% of milk fat content is triacylglycerols, with phospholipids, cholesterol, diacylglycerols, monoacylglycerols and free fatty acids constituting the remainder.

Milk fat contains a multitude of fatty acids: saturated (SFA), monounsaturated (MUFA) and polyunsaturated fatty acids (PUFA) are all present in milk fat. The variety of fatty acids allows the mammary gland to produce triacylglycerols with a range of fluidity, so that the mammary cell can secret the milk fat. The fluidity of the triacylglycerols is increased by use of short and medium chain fatty acids that arise from *de novo* synthesis as well as long-chain unsaturated fatty acids (UFA) [7]. The mammary gland has an additional means of regulating the fluidity of the milk fat via the enzyme Δ^9-desaturase, also named stearoyl-CoA desaturase. The main action of the Δ^9-desaturase is to convert stearic acid (C18:0) to oleic acid, C18:1cis9 [8] and palmitic acid (C16:0) to palmitoleic acid, C16:1 [9].

Milk fat contains a great number of fatty acids. However, many of these are only found in small quantities and there are only 15 or 16 fatty acids present at concentrations at or above 1%. The main SFA, which account for 60-70% of the total fatty acids, are straight-chain molecules which vary in length from 4 to 18 carbon atoms. The principal UFA is oleic acid, around 17-25% of the total. The acid composition of milk fat is affected by factors such as the diet of the animal, the animal breed, and the stage of lactation; for this reason, the proportions of the fatty acids vary throughout the dairying season, in some instances quite markedly.

Health Effects of Oleic Acid

There is ample scientific evidence showing that modulation of dietary fat composition affects blood-lipid concentrations. Regarding oleic acid, or MUFA (MUFA), the most noticeable effects come from studies where the substitution of saturated fat with oleic acid has been tested. Isocaloric replacement of about 5% of energy from SFA by oleic acid (or PUFA) has been estimated to reduce coronary heart disease risk by 20–40%, mainly via LDL-cholesterol reduction [10]. As intakes of saturated fat in many countries are higher than the recommended levels, an increase in the intake of oleic acid may be beneficial as it limits the intake of saturated fat. This can be achieved by changing dietary patterns like using olive oil instead of butter, or by using food technology to modify the fatty acid profile of foods naturally rich in SFA in favor of oleic acid. As oleic acid is the quantitatively most important representative of MUFA in the diet, oleic acid and MUFA intakes and recommendations are virtually the same. According to the recent national surveys, the total intake of oleic acid in adults varies between 12% and 18% of energy (E%), for most countries around 10–12 E%. The intakes are higher in Southern European countries (up to 29%) like Greece, Italy or Spain [11], where the main source of oleic acid is olive oil, and lower in western and northern countries where oleic acid intake comes from meat and mainly meat products. Oleic acid intake recommendations are based on the recommendations for intakes of total fat (around 30% of the total energy), maximum intakes of SFA (10% or less), and minimum and maximum intakes of PUFA (6–10% of the energy). As a consequence, dietary intake of oleic acid should therefore be in the range of 10–15%. While the average intake of fat in the European population is ca. 35% of the total energy, most countries are in the low range of intake (with the exception of the Southern Europe) and should probably increase the consumption of high oleic acid foods like olive oil.

There is a wealth of evidence from epidemiologic and clinical studies suggesting that modifications of dietary fat composition affect the risk of cardiovascular disease [12]. Consumption of n-3 PUFAs, namely eicosapentaenoic acid (EPA) and docosahexaenoic acid (DHA), has several beneficial properties that prevent cardiovascular disease, and inhibit atherosclerosis and thrombosis. Olive oil also is considered a healthy source of fat, and international nutritional guidelines recommend its consumption due to the beneficial cardiovascular effects reported for oleic acid.

Dairy Cow Milk

Bovine milk contains the nutrients needed for growth and development of the calf, and is a resource of lipids, proteins, amino acids, vitamins and minerals. It contains immune-globulins, hormones, growth factors, cytokines, nucleotides, peptides, polyamines, enzymes and other bioactive peptides. The lipids in milk are emulsified in globules coated with membranes. On average, milk contains about 33 g total lipid/L [13]. Triacylglycerols, which account for about 95 % of the lipid fraction, are composed of fatty acids of different lengths (4–24 C-atoms) and saturations.

Each triacylglycerol molecule is built with a fatty acid combination giving the molecule liquid form at body temperature. Other milk lipids are diacylglycerol (about 2% of the lipid fraction), cholesterol (less than 0.5 %), phospholipids (about 1%), and free fatty acids (FFA) accounting to less than 0.5% of total milk lipids. Increased levels of FFA in milk might result in off-flavors in milk and dairy products, and the free volatile short-chain fatty acids contribute to the characteristic flavors of ripened cheese.

The type of fatty acids present in milk fat can influence the flavor and physical properties of dairy products [14]. Milk fat contains substantial amounts of short-chain fatty acids, making it unique compared to other fats. The short-chain fatty acids, particularly butyric acid (C4:0), are important for flavor development in some cheeses and fermented dairy products [15]. It is well documented that spring and summer milk fat from cows on pasture contains more UFA: hence, butter manufactured from this milk fat will be softer. In fact, the ratio of SFA to UFA and the ratio of short to long-chain fatty acids can affect the melting and spreading ability of butter. As indicated in Table 1, medium-chain fatty acids, such as myristic (C14:0) or palmitic (C16:0), and long-chain SFA, such as stearic (C18:0), have the greatest influence on butter hardness, since they have the highest melting point [14].

Table 1. Melting points of the major fatty acids in bovine milk fat

Fatty acid	Melting point (°C)
C4:0 butyric acid	-8
C6:0 caproic acid	10
C8:0 caprylic acid	16
C10:0 capric acid	31.5
C12:0 lauric acid	44
C14:0 myristic acid	58
C16:0 palmitic acid	64
C18:0 stearic acid	69
C18:1 oleic acid	13
C18:2 linoleic acid	-5

Oleic acid (C18:1cis9) is the UFA with the highest concentration in milk accounting to about 8 g/L whole milk, according to Haug et al. [13]. Accordingly, milk and milk products contribute substantially to the dietary intake of oleic acid in many countries. Oleic acid is considered to be favorable for health, as diets with high amounts of MUFA will lower both plasma cholesterol, LDL-cholesterol and triacylglycerol concentrations, as stated by Kris-Etherton et al. [11], and replacement of SFA with cis-UFA reduces risk for coronary artery disease. Oleic acid is more stable to oxidation than the omega 3 and omega-6 fatty acids, and it can partly replace these fatty acids in both triacylglycerols and membrane lipids. A high ratio between oleic acid and PUFA will protect lipids in LDL towards attack from oxidative stressors such as cigarette smoke, ozone and other oxidants. Studies have shown that a diet rich in MUFA/PUFA give better protection against atheromatosis and CVD than a diet rich in PUFA [16]. Milk fat is rich in oleic acid and it has a very high oleic acid/PUFA ratio. A diet rich in milk fat, therefore, may help to increase this ratio in the total dietary fatty acids. Fatty acid composition is quite similar in dairy cows' milk and in sheep's milk, as indicated in Table 2, except for capric acid (C10:0), which is significantly higher in ewe's milk. Both the mammalian species, in fact, are ruminant, and consequently the origin of the different fatty acids content follows a similar metabolic pathway.

The fatty acids of bovine milk are derived from two sources. The first source is fatty acids supplied to the udder by the blood, composed of fatty acids absorbed from the intestine and mobilized from the adipose fat tissue, mainly palmitic acid (16:0), stearic acid (18:0) and longer-chained fatty acids. The second source is derived from circulating blood acetate and butyrate produced during fermentation in the rumen (*de novo* synthesis), and fatty acids up till 14 carbon atoms are synthesised in the udder. Palmitic acid in milk originates from both *de novo* synthesis and from circulating blood. Due to the extensive biohydrogenation of dietary UFA in the rumen, the supply of these fatty acids to the udder is low.

However, in the udder, a desaturation of fatty acids, such as 12:0, 14:0, 16:0 and 18:0, take place, and the products being 12:1, 14:1, 16:1 and 18:1, respectively. The preferred substrate for the desaturating enzyme; delta-9-desaturase, is stearic acid. Therefore, bovine milk is a relatively good source of oleic acid (18:1, cis 9). The udder enzymes cannot make double bonds in omega-3 and omega-6 positions. Consequently, milk content of linoleic acid and alpha-linolenic acid depends on the supply of these to the udder. Conjugated linoleic acid (9c,11t-CLA) in milk originates from two sources.

A small part originates from incomplete biohydrogenation of linoleic acid in the rumen which are absorbed from the small intestine transported to the udder and included in the fat synthesis. Most of the 9c,11t-CLA originates, however, from vaccenic acid which is an intermediate from the biohydrogenation of UFA in the rumen. After absorption and transportation by the blood to the udder, a portion of the vaccenic acid is desaturated by delta-9-desaturase to CLA. There is a close positive correlation between the milk content of vaccenic acid and 9c, 11t-CLA [18].

The effect of feed on milk fat content and fatty acid composition has been previously investigated [19]. There are large variations in fat synthesis in the udder, and the fat content and fatty acid composition are the most modifiable of the main components in milk. There are seasonal variations for the major fatty acids [20].

Milk from grazing dairy cows contain significantly higher proportions of oleic acid than milk produced on traditional indoor feeding composed of concentrates and conserved roughages.

Typically, CLA content in milk produced on pasture is at least twice of that obtained by indoor feeding. Moreover, the proportion of α-linolenic acid increases more than linoleic acid, resulting in a lower ratio between omega-6 and omega-3 fatty acids. Milk fat from cows fed an indoor diet consisting of conserved grass and concentrate have a ratio between omega-6 and omega-3 fatty acids of about 4:1; but in summer when the cows are out on pasture and have a high intake of grass the ratio may be reduced to about 2:1.

These positive effects of pasture on the fatty acid composition of milk are mainly attributed to the high content of PUFA, especially α-linolenic acid, in grasses at early stages of maturity [21].

Table 2. Fatty acid composition of dairy cows and ewe's milk

Fatty acid (% of total fatty acid)	Dairy cow	Ewe
C4:0	3.61	3.17
C6:0	1.97	2.22
C8:0	1.31	2.22
C10:0	2.62	6.35
C12:0	2.95	3.81
C14:0	11.15	10.48
C16:0	28.85	25.71
C16:1	2.62	2.06
C18:0	13.11	14.29
C18:1	27.54	24.76
C18:2	2.62	2.86
C18:3	1.64	2.06

Source: Adapted from Vera [17].

Ewe's Milk

The average composition of ewe's milk includes 7.1% fat, 5.7% protein, 4.8% lactose, 9% ash and 11.5% non-fatty solids [22]. Fat and protein contents are high in early lactation, decrease during the lactation peak and then increase again as the milk yield diminishes. Lactose content shows little variation, as its synthesis determines the rate of daily milk production, but the negative relation between yield and fat and protein contents generally holds, regardless of whether milk yield is a result of genetics, variation among animals or selection processes [17].

About 50% of milk fatty acids are synthesized in the mammary gland from blood acetate and β-hydroxybutyrate by the action of acetil-CoA carboxylase and fatty acid synthetase [23]. The remaining part comes both from the diet (about 40-45%) and from the mobilization of body lipid reserves of the animal, in different proportions according to the lactation stage. Such percentages may vary significantly with dietary manipulations. Since the glandular cells of the udder do not have the enzymatic pool necessary to elongate the carbon chain from 16 to 18 atoms, the neo-synthesized fatty acids are made up of short- and medium-chain fatty acids only, ranging from 4 to 16 carbon atoms. All milk fat acids with chain lengths equal or longer than 18 carbon atoms must be provided from the diet, which strongly influences the content of long-chain fatty acids in milk.

Milk fat fluidity is affected by the desaturase activity of the mammary gland. In fact, the Δ^9-desaturase enzyme introduces a double bond in position 9 of the carbon chain, and for this reason a large amount of stearic acid (C18:0) is desaturated into oleic acid (C18:1cis9), reaching a ratio between the two fatty acids that optimizes the physical-chemical characteristics of milk.

Fats can be added to sheep diets with the aim to meet the high energy requirements of the animals during the first phase of lactation, to increase milk fat content and to modify the milk fatty acid profile. The role of fats in ruminant nutrition can be basically summarized as follows [24]:

1. to increase gross energy intake when dry matter contribution is limited;
2. to increase net energy, optimizing the loss of heat;
3. to modify milk fat composition with the transferring of dietary fatty acids to milk;
4. to partially replace the highly fermentable carbohydrates.

All depend on several factors such as digestibility of fats added to the diet, influence of this supplementation on voluntary intake and on digestibility of other components in the diet, and obviously the kind of fat added. The amount of fat added to the diet may affect dry matter intake: in fact, if dietary fat supplementation exceeds 6%, dry matter intake can decrease. Lipids, moreover, can modify the ratio between volatile fatty acids in the rumen, perhaps as a consequence of a negative effect on cellulolytic bacteria rather than amylolytic microbes, which induces an increase in propionate [25]. To avoid negative effects on milk fat composition, several methods of fat rumen protection have been used. In dairy ewes, calcium soaps of fatty acids have been tested as a source of lipid supplementation. The general effect of calcium salt supplementation was an increase in the UFA and a decrease of SFA in milk [26]. The inclusion of calcium soaps of olive oil in the ewe diets induced an increase of total octadecenoic acid content in milk [27], but it is not clear if this increase is related only to the oleic acid content (C18:1cis9) or also to other isomers of C18:1 that originate from ruminal biohydrogenation processes. An *in vitro* study [28] on the ruminal biohydrogenation of oleic acid seemed to confirm the second hypothesis.

Many studies have confirmed that it is possible and relatively easy, rapid and efficacious to affect milk fat quality in ruminant species by changing the composition of the diet at the farm level and in a reversible manner [29]. It is relatively easy to affect the milk fat quality at the farm level simply by changing the forage/concentrate ratio. There is a great difference in diets based on forage only and diets with concentrates. With 100% forage, the quality of milk fat is surely better, with a lower concentration of lauric acid (C12:0) and palmitic acid (C16:0), known as atherogenic and thrombogenic acids [30], and higher amounts of UFA, such as oleic acid (C18:1) and Conjugated Linoleic Acid (CLA), the beneficial ones. Also, the nature of forage is an important factor: using forages from a high-hill pasture it is possible to obtain a higher amount of oleic acid in milk; whereas the concentration of myristic acid (C14:0), the most harmful fatty acid is not different between pastures, but much lower compared to that in milk of cows fed concentrates.

Mares Milk

In the past centuries, mare's milk has been one of the most important foods for the nomadic populations of central Asia and the former Soviet Union, where it is still used mainly for the manufacturing of a lactic-alcoholic beverage, named Koumiss. Recently, there has been an increasing interest in the consumption of mare's milk in the western countries as well, in particular in Germany, where mare's milk production and commerce for human use is regulated by law. The benefits of mare's milk for human health regard mainly the treatment of gastrointestinal ulcers and chronic hepatitis [31].

In horses, as in other mammals, milk fat has a double origin: circulating fatty acids transported in lipoproteins, and fatty acids synthesised in the udder. Circulating fatty acids mainly come from dietary 16 and 18 carbon fatty acids, which are adsorbed in the small intestine in horses. As in other monogastric species, they are not metabolized before adsorption, contrary to ruminants, in which fatty acids are transformed prior to absorption by isomerisation and hydrogenation, and to a lower extent synthesized, in the rumen [32]. In the large intestine, fatty acids metabolism occurs due to bacterial activity.

Mare milk composition is very different from ruminant milk composition, because of the above mentioned extensive hydrogenation of dietary fatty acids in ruminants. The main difference in composition with other monogastric species is that horses consume grass, which is rich in linolenic acid (C18:3), whereas feedstuffs consumed by non-herbivorous monogastric species are most often poor in this fatty acid. Compared to human milk, mare's milk is richer in short- and medium-chain fatty acids and in linolenic acid, while it is poorer in oleic acid (Table 3).

Mare's milk is rich in short- and medium-chain fatty acids; the proportion of short- and medium-chain fatty acids is higher than that of cow or human milk fat. A specificity of mare's milk fat is the very low concentration of stearic acid (C18:0), ranging from 0.3 to 3.0% of total fatty acids: among mammals, only seals have such a low concentration.

Table 3. Fatty acid composition in mare and human milk (% of total fatty acids)

Fatty acid	Mare's milk	Human milk
C4:0	0.6	---
C6:0	0.7	traces
C8:0	3.1	traces
C10:0	7.2	1.3
C12:0	7.6	3.1
C14:0	7.5	5.1
C14:1	0.7	0.5
C15:0	0.4	traces
C15:1	0.4	traces
C16:0	20.8	20.2
C16:1	5.6	5.7
C17:0	0.4	0.6
C17:1	0.4	0.3
C18:0	1.2	6.0
C18:1	19.8	46.4
C18:2	11.3	13.0
C18:3	11.9	1.4

Source: Adapted from Doreau and Martuzzi [32].

The reason for a such low content of stearic acid in mare's milk can be attributed to the low content of this fatty acid in the diet for horses, and also to a possible high activity of Δ^9-desaturase, which converts stearic acid into oleic acid (C18:1cis9) in the udder. This second hypothesis is enhanced by a high content also in palmitoleic acid (C16:1cis7), converted by the same enzyme [33]. Mare's milk is also rich in PUFA: the important content of linoleic acid (C18:2cis6) is due to the presence of this fatty acid in the diet, representing the major fatty acid in cereals and in their by-products. Another specificity of mare's milk is the high content in α-linolenic acid (C18:3cis3); in milk from animals which are both monogastric and herbivorous, as are mares, the reason for the high content of this fatty acid is connected to the diet, normally based on grass, rich in α-linolenic acid, and is also due to the specific digestive physiology of being monogastric, in which the hydrogenation of fatty acids does not occur before absorption.

Donkey's Milk

Donkey's milk has recently aroused scientific interest, above all among pediatric allergologists and nutritionists. Clinical studies have demonstrated that donkey milk may be considered a good replacement for dairy cow's milk in feeding children with a severe Ig-E mediated cow's milk protein allergy, when human milk cannot be given [34, 35]. For these patients, donkey milk is only not useful, but also safer compared with milk obtained from other mammalian species [36], due to the high similarity with human milk, especially considering protein fractions content [37; 38].

Donkey milk shows a lower fat content compared to human milk, presenting a reduced energetic value for this reason [39]. The knowledge of the composition of milk lipids is fundamental in order to explain the physical properties, nutritive characteristics and biosynthetic pathways of milk fat. The large number of fatty acids present in the lipid fraction of milk makes it one of the most complex naturally occurring fats. SFA are the most represented classes in donkey's milk compared to MUFA and PUFA, even though a wide variability can be observed in the data available in literature (Table 4), most likely related to dietary and/or body condition differences [40].

The equine large intestine (cecum, ventral colon, distal colon and small colon) mucosa exhibits significant differences in relative rate of transport of volatile fatty acids compared with rumen mucosa [41]. Moreover, it is known that *ex novo* synthesis of part of C18 fatty acids occurs from the products of the large bowel fermentation, differently from 4 to 14 carbon fatty acids and part of 16 carbon fatty acids. It is really very interesting to mention the high content determined for the PUFA in donkey's milk, while a conjugated linoleic acid group has not been detected, according to data obtained on mare's milk [42]. The average amount of oleic acid content in donkey's milk is around 15 g/100 g of total fatty acids, according to Chiofalo et al. [43]. Another study [44] confirmed that oleic acid is the most abundant MUFA in donkey's milk, with a content of 19.8 g/100 g of total fatty acids. Donkey's milk significantly differs from human milk and dairy cows milk for its lower total lipid content (0.94% vs 3.8% and 3.6%, respectively) which is reflected in its lower caloric content [45].

Table 4. Range of variation of donkey's milk fatty acid composition

Fatty acids (g/100 g)	Range
Saturated Fatty Acids (SFA)	46.7 – 67. 6
Monounsaturated Fatty Acids (MUFA)	15.3 – 28.2
Polyunsaturated Fatty Acids (PUFA)	16.0 – 22.8
Ratio Unsaturated/Saturated	0.48 – 1.07

Source: Adapted from Salimei and Chiofalo [40].

Donkey's milk contains few SFA (5.46 g/L) and few MUFA (1.96 g/L), whereas the PUFA content of donkey's milk is slightly higher than in dairy cows milk but lower than in human milk. Among the SFA contained in donkey's milk medium-chain fatty acids predominate (Table 5). Considering the nutritional characteristics of the donkey milk lipid profile, medium-chain fatty acids undergo preferential intestinal absorption versus long-chain fatty acids, which are present in very small quantities in donkey milk. Within a well-balanced and integrated diet, donkey's milk is a good source of essential fatty acids; this category of fatty acids is very important in the diet of patients with cow's milk protein allergies (CMPA), especially if affected by multiple food allergies. These subjects are in fact at risk of developing a deficiency in essential fatty acids and particularly in PUFA n-3, which are absolutely necessary for adequate growth, neurological development and cardiovascular health [46]. Donkey's milk shows an high content of both linoleic (C18:2) and linolenic (C18:3) acids, respectively 9.0 g/100 g and 5.1 g/100 g of total fatty acids, when compared with ruminant species milks, in which the contents of the above mentioned PUFAs are always lower.

Table 5. Comparison of fatty acid composition between donkey's milk, human milk and dairy cow milk

Donkey's milk		Human milk	Dairy cows milk
Total fat (%)	0.94	3.80	3.60
SFA (mg/100 ml)	546	1520	2580
MUFA (mg/100 ml)	196	1690	920
PUFA (mg/100 ml)	169	578	131
PUFA n -6 (mg/100 ml)	94.0	509	103
PUFA n – 3 (mg/100 ml)	74.6	68.9	28.0

Source: Adapted from Gastaldi et al. [45].

Table 6. Total fatty acids composition of triacylglycerols fraction in cow, ewe and donkey milk (mean values±s.d.)

Fatty acid (% of total fatty acids)	Cow	Ewe	Donkey
C4:0	6.0±0.8	10.3±1.5	2.1±1.1
C6:0	2.9±0.6	3.4±0.7	0.5±0.5
C8:0	1.7±0.2	2.5±0.5	4.1±0.5
C10:0	3.4±0.6	6.1±2.1	9.5±3.4
C12:0	3.9±0.9	3.0±0.7	8.9±5.1
C14.0	13.1±1.2	7.0±0.4	7.5±3.3
C16:0	31.6±2.6	19.8±0.7	20.9±1.3
C16:1n-7	1.8±0.3	0.7±0.1	5.4±2.0
C18:0	6.6±1.4	11.8±1.2	1.5±0.8
C18:1n-9	19.2±4.3	22.8±2.1	19.8±4.7
C18:2n-6	1.9±0.2	2.7±1.5	9.0±3.1
C18:3n-3	0.2±0.1	0.8±0.2	5.1±3.9
Others	7.7	9.2	5.7
n-6/n-3 ratio	9.5	3.4	1.8

Source: Adapted from Blasi et al. [44].

In Table 6, the total fatty acids composition of the triacylglycerols fraction in dairy cows milk, ewe's milk and donkey milk is shown. The n-6/n-3 ratio is a very interesting index from a nutritional point of view. Donkey milk has the lowest n-6/n-3 ratio (see Table 6), from 2 to 6 times lower than the other mammalian species. Therefore, because the recommended n-6/n-

3 fatty acids ratio is estimated to be 4/1, while the current dietary n-6/n-3 ratio is about 10/1 [47; 48], novel dietary sources with a low n-6/n-3 PUFA ratio, as donkey milk could be, are in high demand for improving human nutrition.

CONCLUSION

The nutritional quality of both mare's milk and donkey's milk in human health as possible replacements for dairy cows milk can be mainly attributed to their nitrogen fractions. Considering fat content, the main interest in using mare's milk for human nutrition is related to the high proportion of α-linolenic acid, which is important in the prevention of cardiovascular diseases, and in the synthesis of prostaglandins and other long-chain fatty acids which have different actions on the immune system, and for infant growth. Furthermore, mare's milk also has anti-inflammatory properties. Donkey's milk has a notable nutritional significance because of its lower amount of SFA and because it represents an important dietary source of Essential Fatty Acids, widely known for the fundamental functions in the human metabolism.

Humans are the only animals who take milk as adults. Interestingly, the nutritional requirements of humans change with age and physiology: they are not the same after birth, in childhood, during adolescence, in the adult period, during pregnancy, lactation, menopause, etc. For this reason, if we look for optimal nutrition, it makes sense to adapt the composition of standard cow's milk to the nutritional requirements and physiology of the different age periods. Moreover, the composition of cow's milk maybe ideal for calves, but not necessarily for humans. In fact, when breastfeeding is not possible, the infant formula available on the market tries to imitate the human breast milk, adapting their compositions to the nutritional requirements at the different early ages: 0–6 months, 6–12 months, and 1–3 years. If these concepts are extended to the adult age or the different physiological statuses, modifying the composition of standard cow's milk may have many advantages. The substitution of the saturated milk fat by oleic acid and omega 3 has shown cardiovascular benefits.

Milk is an everyday used drink which has been used as a carrier to try to increase the intake of certain nutrients (vitamins, minerals, protein, fats) and/or bioactives (phytosterols, probiotics, antioxidants, isoflavones, etc.). We will see more types of enriched or modified dairy drinks aimed to promote organ health and function and adapted to the different ages and physiology, an area of research that will certainly develop in the future.

REFERENCES

[1] A.J.B. Lucas, B. Coulon, P. Grolier, B. Marin and E. Rock, *Indicators of Milk and Beef Quality,*. Wageningen, Holland (2005).

[2] Y.M. Pacheco, B. Bermúdez, S. López, R. Abia, J. Villar and F.J.G. Muriana, F.J.G., *Amer. J. Clin. Nutrit.* 84, 342 (2006).

[3] P.J. Huth, D.B. Di Rienzo and G.D. Miller, *J. Dairy Sci.* 89, 1207 (2006).

[4] S. Renaud, M. de Lorgeril, J. Delaye, J. Guidollet, F. Jacquard, N. Mamelle, J.L. Martin, I. Monjaud, P. Salen and P. Toubol, *Amer. J. Clin. Nutrit.* 61, 1360S (1995).

[5] R.W. Owen, R. Haubner, G. Würtele, E. Hull, B. Spiegelhalder and H. Bartsch, *Europ. J. Cancer Prev.*, 13, 319 (2004).
[6] M.A. McGuire and D.E. Bauman, *Encyclopedia of Dairy Science*, Academic Press (2003).
[7] D.E. Bauman, and J.M. Griinari, *Livest. Prod. Sci.* 70, 15 (2001).
[8] R.R. Grummer, *J. Dairy Sci.* 74, 3244 (1991).
[9] G. Savoini, A. Agazzi, G. Invernizzi, D. Cattaneo, L. Pinotti, and A. Baldi, *Small Rumin. Res.* 88, 135 (2010).
[10] E. Lòpez-Huertas, *Pharm. Res.*, 61, 200 (2010).
[11] P.M. Kris-Etherton, T.A. Pearson, Y. Wan, R.L. Hargrove, K. Moriarty, V. Fishell, and T.D. Etherton, *Amer. J. Clin. Nutrit.*, 70, 1009 (1999).
[12] J.J. Carrero, L. Barò, J. Fonollà, M. Gonzàlez-Santiago, A. Martıìnez-Férez, R. Castillo, J. Jiménez, J.J. Boza, and E. Lòpez-Huertas, *Nutrit.*, 20, 521 (2004).
[13] Haug, A.T. Høstmark, and O.M. Harstad, *Lipids Health Diseases* 6, 25 (2007).
[14] R.J. Baer, *J. Dairy Sci.* 54, 383 (1991).
[15] P. Polidori, G.L. Maggi, V.M. Moretti, F. Valfrè and P. Navarotto, *Anim. Prod.* 57, 319 (1993).
[16] R.P. Mensink, P.L. Zock, A.D. Kester and M.B Katan, *Amer. J. Clin. Nutrit.* 77, 1146 (2003).
[17] R.R. Vera, C. Aguilar and R. Lira, *Ciencia Investig. Agr.* 36, 307 (2009).
[18] M. Antongiovanni, A. Buccioni, F. Petacchi, P. Secchiari, M. Mele and A. Serra, *Ital. J. Anim. Sci.* 2, 3 (2003)
[19] J.A. Bell, J.M Griinari and J.J. Kennelly, *J. Dairy Sci.* 89, 733 (2006).
[20] M. Ledoux, J-M. Chardigny, M. Darbois, Y. Soustre, J-L Sebedio and L. Laloux, *J. Food Comp. Anal.* 18, 409 (2005).
[21] A.T. Ward, K.M. Wittenberg, H.M. Froebe, R. Przybylski and L. Malcomson, *J. Dairy Sci.* 86, 1742 (20039.
[22] T.T. Treacher and G. Caja, (2002). *Sheep Nutrition*, Wallingford, U.K. (2002).
[23] M. Mele, A. Buccioni, A. Serra, M. Antongiovanni and P. Secchiari, *Dairy Goats Feeding and Nutrition*, Oxfordshire, UK (2008).
[24] D.L. Palmquist, *J. Nutr.* 124, 1377S (1994).
[25] M. Doreau, Y. Chilliard, D. Bauchart and B. Michalet-Doreau, *Annal. Zootech.* 40, 19 (1991).
[26] Baldi, F. Cheli, C. Corino, V. Dell'Orto and F. Polidori, *Small Rumin. Res.* 6, 303 (1992).
[27] M. Mele, A. Buccioni and A. Serra, *Ital. J. Anim. Sci.* 4(Suppl. 1), 53 (2005).
[28] E.E. Mosley, G.L. Powell, M.B. Riley and T.C. Jenkins, *J. Lipid Res.* 43, 290 (2002)
[29] Y. Chilliard, A. Ferlay and M. Doreau, *Livest. Prod. Sci.* 70, 31 (2001).
[30] T.L.V. Ulbricht and D.A.T. Southgate, *Lancet 338*, 985 (1991).
[31] F. Martuzzi and M. Doreau, *Nutrition and Feeding of the Broodmare*, Wageningen, Holland (2006).
[32] M. Doreau and F. Martuzzi, *Nutrition and Feeding of the Broodmare,* Wageningen, Holland (2006).
[33] M. Doreau, S. Boulot and Y. Chillard, *J. Dairy Res.* 60, 457 (1993).
[34] Carroccio, F. Cavataio, G. Montalto, G. D'Amico, L. Alabrese and G. Iacono, *Clin. Exp. Allergy* 30, 1597 (2000).

[35] G. Monti, E. Bertino, M.C. Muratore, A. Coscia, F. Cresi, L. Silvestro, C. Fabris, D. Fortunato, M.G. Giuffrida and A. Conti, *Pediatr. Allergy Immunol.* 18, 258 (2007).

[36] P. Polidori, D. Beghelli, P. Mariani and S. Vincenzetti, *Ital. J. Anim. Sci.* 8 (Suppl. 2), 677 (2009).

[37] E. Salimei, F. Fantuz, R. Coppola, B. Chiofalo, P. Polidori and G. Varisco, *Anim. Res.* 53, 67 (2004).

[38] S. Vincenzetti, P. Polidori, P. Mariani, N. Cammertoni, F. Fantuz and A. Vita, *Food Chem.* 106, 640 (2008).

[39] P. Dugo, T. Kumm, M. Lo Presti, B. Chiofalo, E. Salimei, A. Fazio, A. Cotroneo and L. Mondello, *J. Separation Sci.* 28, 1023 (2005).

[40] E. Salimei and B. Chiofalo, *Nutrition and Feeding of the Broodmare*, Wageningen, Holland (2006).

[41] P.J. Van Soest, *Nutritional ecology of the ruminant*, Comstock Publishing Associates (1994).

[42] M. Malacarne, F. Martuzzi, A. Summer and P. Mariani, *Int. Dairy J.* 12, 869 (2002).

[43] B. Chiofalo, C. Drogoul and E. Salimei, *Nutrition and Feeding of the Broodmare*, Wageningen, Holland (2006).

[44] F. Blasi, D. Montesano, M. De Angelis, A. Maurizi, F. Ventura, L. Cossignani, M.S. Simonetti and P. Damiani, *J. Food Comp. Anal.* 21, 1 (2008).

[45] D. Gastaldi, E. Bertino, G. Monti, C. Baro, C. Fabris, A. Lezo, C. Medana, C. Baiocchi, M. Mussap, F. Galvano and A. Conti, *Frontiers in Bioscience E2*, 1, 537 (2010).

[46] L. Aldamiz-Echevarria, A. Bilbao, F. Andrade, J. Elorz, J.A. Prieto and J. Rodriguez-Soriano, *Acta Paediatr.* 97, 1572 (2008).

[47] P.M. Kris-Etherton, D.S. Taylor, S. Yu-Poth, P. Huth, K. Moriarty, V. Fishell, R.L. Hargrove, G. Zhao and T.D. Etherton, *Amer. J. Clin. Nutr.* 71, 179S (2000).

[48] P.M. Kris-Etherton, *Circulation* 100, 1253 (1999).

In: Oleic Acid: Dietary Sources, Functions and Health Benefits ISBN: 978-1-62618-332-2
Editor: Luciano Paulino Silva

Chapter 8

OLEIC ACID AND BLOOD BRAIN BARRIER

Hyung Soo Han*

Department of Physiology, Brain Science and Engineering Institute,
Kyungpook National University School of Medicine, Daegu, Korea

ABSTRACT

Blood brain barrier (BBB) acts as a physical and functional barrier between blood circulation and brain parenchyma. Since BBB maintains the cerebral microenvironment in a strict manner, it is essential in protecting the brain from the exogenous deleterious substances such as bacteria, viruses, or toxins in the pathological conditions. But tight restriction of BBB poses a significant obstacle to the delivery of most drugs into the brain. As a result, drug therapy has rarely been available to treat brain tumors, cancer metastases to brain and other brain diseases and therefore delivery system to overcome BBB is crucial to the central nervous system (CNS) drug developers. Fatty acids are the major constituents of lipids in our body. The nature of the fatty acids in the cell membrane contributes greatly to the fluidity of the membrane and it is believed that fatty acid plays a role in regulating the activity of membrane proteins. In addition, membrane phospholipids are a source of second messenger molecules, such as diacylglycerol, phosphatidic acid, inositol 1,4,5-triphosphate, ceramide and arachidonic acid that are involved in the regulation of signaling pathways from the membrane to the cytosol or nucleus. Some fatty acids are substrates for the synthesis of bioactive molecules such as prostaglandins, thomboxanes and leukotrienes. Oleic acid is one of the major constituents of unsaturated fatty acids of human fat. In experimental model of cerebral fat embolism induced by intra-arterially administered oleic acid, brain edema appeared in the hyperacute stage. This transient edema formation is related with reversible dysfunction of BBB. The mechanisms of enhanced BBB permeability by oleic acid have been investigated. Oleic acid induced BBB alteration was visualized by microscopic observation of extravasated Evans blue or FITC-labeled dextran in addition to gadolinium-enhanced T1 magnetic resonance imaging (MRI). Dysfunction of BBB was quantified by measuring Evans blue concentration in the brain tissue or counting endothelial barrier antigen (EBA) immunoreactivity. Oleic acid induced inflammation in the cerebral vessels and surrounding brain tissues was observed. Induction of intercellular

* E-mail address: hshan@knu.ac.kr

adhesion molecule-1 (ICAM-1) in the vascular endothelial cells and migration of neutrophils into the brain parenchyma were observed. In conclusion, inflammation might be one of the major contributing factors of oleic acid induced BBB disruption.

INTRODUCTION

Oleic acid is one of the major constituents of unsaturated fatty acids of human fat. Fatty acids can regulate ion channel activation directly in certain cells. For example, arachidonic acid and other fatty acids activate potassium ion channels in smooth muscle cells and calcium channels in cardiac sarcolemmal vesicles [1, 2]. Long-chain fatty acids trigger the release of insulin by β-cells through the activation of a G-protein-coupled receptor [3]. Total fatty acid concentration in the blood is roughly 0.4 - 0.5 mM. However, because 99% of circulating fatty acids are bound to albumin, the amount of unbound free fatty acid which is an available form to the brain and other tissues is quite small. At physiological levels, the concentration of unbound free fatty acids is about 7.5 nM [4]. It is reported that high olive oil consumers have higher plasma oleic acid levels and are associated with lower incidence of stroke in older subjects [5]. In spite of the beneficial effects, acute abnormally high oleic acid level in the blood can cause detrimental effect to the blood vessels and adjacent tissues as it is usually observed in fat embolism.

Fat embolism is caused by circulating fat emboli in the blood stream and has been reported to occur in several clinical conditions including bone marrow transplant, pancreatitis, fatty liver, liposuction and others. But fat embolism is associated primarily with osseous fractures of long bones, and the clinical fat embolism syndrome has been reported in up to 11% of these cases. That is, fat emboli occur in almost 90% of all patients with severe injuries to bones but only 10% of these are symptomatic [6]. Symptomatic or clinical fat embolism typically involves multi-organ systems such as lung, brain, blood, skin, heart, eye, kidney systems. Once fat is leaked from the damaged tissue into the blood circulation, it forms emboli and occludes the pulmonary microvasculature first. Depending on the size of fat globules, smaller globules may traverse the pulmonary microvasculature and reach the systemic blood circulation, leading to diverse systemic symptoms or manifestations of fat embolism depending on the organ where its microcirculatory bed is blocked. Symptoms of fat embolism include a spectrum of normal or subclinical, mild to devastating features. When fat emboli occlude the small vessels supplying the important organs or tissues, embolism leads to injury to the tissues. For example, acute lung injury and acute respiratory distress syndrome may result from fat emboli occluding pulmonary capillaries and biochemical alterations that produce lung injury that directly damages the pulmonary capillary endothelium. Systemic embolization is observed most commonly in the brain and kidneys. But clinical cerebral fat embolism is a rare, yet potentially detrimental, complication of bone fractures. Neurological signs from the cerebral fat embolism syndrome include paresis or paralysis, convulsions, delirium, confusion or seizure, and stupor, which can lead to coma or brain death [7]. Depending on the severity of embolism and affected brain area, neurological symptoms vary from confusion with diminished Glasgow coma scale to encephalopathy with coma and seizures. The diagnosis in clinical field may be difficult as cerebral fat embolism may be masked by other combining disease conditions. Pathologically diffuse encephalopathy, localized brain hemorrhages, localized cerebral edema, and white matter changes have been seen in patients with cerebral fat embolism. MRI may be necessary to show the characteristic cerebral lesions of the acute state of fat embolism while computed tomography (CT) scan appears normal.

The classic mechanism of fat embolism is explained by mechanical and biochemical theories [7]. The mechanical theory postulates that fat particles from injured adipose tissue enter the blood circulation and obstruct the vessels leading to ischemic events of the tissues. According to the biochemical hypothesis, local hydrolysis of triglyceride emboli by adjacent cellular lipase produces free fatty acids such as oleic acid. Newly formed oleic acid together with excessive mobilization fraction from peripheral adipose tissue result in the toxic concentration of oleic acids in the blood stream leading to damage of nearby tissue. Since oleic acid and other fatty acids have a strong affinity for calcium ions and form calcium soaps, immobilization of the calcium ions results in a loss of intercellular cohesion and a consequent disruption of the endothelial surface of the capillaries. In experimental fat embolism, free oleic acid is known to be more toxic than neutral fat in terms of tissue damage [8]. Clinical symptoms are depending how the affected tissues are damaged. For example in the pulmonary system, oleic acid or related toxic substances make injury to the capillary endothelium first and then increases vascular permeability, producing alveolar hemorrhage, edema, and inactivate the surfactant molecules. Ultimately, these pulmonary alterations lead to respiratory failure, resulting in acute lung injury and acute respiratory distress syndrome. As fat accumulates in the pulmonary microcirculation and lipoprotein lipase liberates free oleic acid, disseminated intravascular coagulation and platelet aggregation further compound capillary disruption and systemic inflammation. Even though there are some papers suggesting the mechanism of fat embolism in the pulmonary system, there are few data sufficient to explain the details of cerebral fat embolism yet. Therefore more recent data with the perspectives on blood brain barrier (BBB) disruption, inflammation, free radical formation and hemodynamic changes from cerebral fat embolism will be mentioned in this chapter.

BLOOD BRAIN BARRIER (BBB)

BBB acts as the physical and functional barrier between the blood circulation and brain parenchyma. The tight controlled passage of substances across the BBB provides a physical selective barrier between the CNS and the peripheral blood circulation and limits passage of certain substances while it facilitates transport of other substances. Since BBB is a protective barrier against potential harmful substances by excluding and effluxing waste products back into the periphery, it maintains the cerebral microenvironment in a strict manner [9].

The special properties of the BBB were first observed in the late 19th century by the German bacteriologist Paul Ehrlich. He found that when he injected dyes into the blood stream they leaked out of capillaries in most regions of the body to stain the surrounding tissues. However, brain remained unstained. Ehrlich wrongly surmised that the brain had a low affinity for the dyes. It was his student, Edwin Goldman, who did the other half of the experiment and realized the truth of what was going on. Goldman injected a dye into the cerebrospinal fluid that surrounds the brain and observed that it stained the brain, but nothing else. Goldmann correctly concluded that the dye was unable to cross the specialized walls of brain capillaries and coined the term BBB for the first time. BBB remained highly debated until the 1960s when The debate as to whether astrocytic end feet or capillary endothelium comprise the BBB was laid to rest by electron microscopic cytochemical studies performed in the late 1960s by Reese and Karnovsky, and later by Brightman and colleagues.

The brain microvascular system representing the characteristics of BBB is formed by a complex cellular system of endothelial cells, non-vascular cells, such as astroglia, pericytes, perivascular macrophages, microglia and a basal lamina or basement membrane [9]. Same basement membrane is shared by both the endothelial cell and the pericyte. Astrocytes project their end feet or foot processes tightly to the endothelium, influencing and conserving the barrier function of endothelial cells. The astrocytic foot processes interact with most of the surface area of the basement membrane. In addition to astrocytes, several other cell types exist in the perivascular space, and they also have the potential to affect changes in endothelial cell function. Microglia, the resident phagocytes of the brain, also extends the processes into the perivascular basement membrane where they can directly contact endothelial cells. Like their macrophage counterparts in peripheral tissues, microglia can release lots of active cytokines and enzymes [10]. Perivascular macrophages, which are functionally and cytologically distinct from microglia, reside within the basement membrane circumscribing the endothelium and are involved in scavenger functions, the secretion of vasoactive cytokines and eicosanoids, and the release of matrix-degrading proteases [11]. Endothelial cells are embedded in the basal lamina together with pericytes and perivascular macrophages. Pericytes are characterized as contractile cells that surround the brain capillaries with long processes, and are believed to play a role in controlling the growth of endothelial cells. Due to their close contact with the endothelial cells, pericytes may influence the integrity of the capillaries and conserve the barrier function. Pericytes additionally limit the transport by the ability to phagocytose compounds which have crossed the endothelial barrier, as observed in a healthy BBB [12].

In most tissues, the capillary endothelial cells are separated by small gaps, allowing solutes to pass freely into and out of the blood and into the surrounding tissues (paracellular route). In addition, most of the endothelial cells in the body have a pore structure called fenestration providing the transcellular route for the solute passage. The lumen of the cerebral capillaries is covered by endothelial cells in which the functional and morphological basis of the BBB resides. CNS endothelium exhibits various functional and morphological differences in comparison with endothelial cells derived from peripheral organs. In the brain, capillaries are a continuous layer of endothelial cells with highly tight adhesion structures between cells and endothelial cells are lack of fenestrations. These junctional complexes between two adjacent cerebral endothelial cells are called tight junction and adherens junction and composed of a multitude of proteins [9]. Tight junctions locate in the upper lateral side of the endothelial cells near the lumen-side and adherens junctions are in the mid to lower lateral side of the endothelial cells. These junctional complexes hinder paracellular transport of hydrophilic compounds across the cerebral endothelium. The absence of fenestrations in the endothelial plasma membrane and the presence of high densities of mitochondria in the cytosol are other prominent morphological features of the CNS endothelium. In addition to junctional complexes, specialized transport systems are present within the membranes of cerebral endothelial cells, which further contribute to the selective transport of solutes into and out of the CNS [13, 14]. In summary both adherens junctions and tight junctions restrict the paracellular transport across the BBB. This low paracellular permeability of solutes are demonstrated as high transendothelial electrical resistance of 1000–3000 $\Omega\ cm^2$ across the BBB in the experimental settings.

Tight junctions are composed of major proteins and accessory proteins [15]. 3 major proteins include junctional adhesion molecule (JAM-1), occludins and claudins. JAM-1, a single transmembrane protein, provides the initial attachment between adjacent endothelial cell membranes through homophilic interactions of the extracellular domains. Occludins and claudins interact between two adjacent endothelial cells and form non-covalent bonds through their extracellular loop domains. Claudins provide the primary seal for the tight junctions and occludins support this seal. Accessory proteins including ZO-1, -2 and -3 located in the cytoplasm and are associated with tight junctions by providing a crosslink to cytoskeletal proteins such as actin. In addition to seal formation, tight junctions provide asymmetric distribution of membrane components, such as transport proteins, by separating the apical and basolateral cell membranes. Due to the presence of tight junctions and specific transport mechanisms on both the luminal and abluminal membranes, BBB forces substances to move transcellularly rather than paracellularly. The endothelial cells and astrocytes in the BBB provide a metabolic barrier for the brain against permeable substances. Both intracellular and extracellular enzymes contribute to the metabolic barrier and protect the brain from potentially harmful compounds by inactivating them into polar compounds, thus preventing them from further crossing into the brain [16].

Since brain is capable of synthesizing only a few non-essential fatty acids, both essential fatty acids and even some non-essential fatty acids must be transported to the brain from the blood [17]. But fatty acids do not readily cross the BBB as components of phospholipids or triglycerides. Fatty acid saturation and chain length appear to play an important role in the transport of fatty acids across the BBB. However, the precise mechanism of transport of fatty acids into the brain is unknown. Fatty acid transport into the brain is more complicated due to the BBB. Because 99% of circulating fatty acids are bound to albumin, the amount of fatty acid available to the brain and other tissues is quite small. Proteins are unable to cross the BBB, thus fatty acids must dissociate from the plasma protein and transfer to the BBB membrane in order to enter into the brain. The presence of transcellular simple diffusion and fatty acid transport protein-mediated movement of fatty acids across the BBB on both the luminal and abluminal membranes of the brain capillaries is reported [18, 19].

In case of fat embolism, the physiologic fatty acid transport system is not working properly and abnormally high fatty acid level results in injury to the endothelial cells. In experimental cerebral fat embolism induced by intra-arterially administered oleic acid, brain edema appeared in the hyperacute stage. This edema formation related with reversible dysfunction of BBB or enhanced BBB permeability has been demonstrated in animal model [20]. MRI yields considerable amounts of useful information to researchers in BBB studies. Currently, high-field MR machines for small animals are able to obtain good quality brain images, even in mice. However, this MR machine is very expensive. On the other hand, clinical MR machines for humans are not suitable for studies of small animals. Owing to the recent development of software and surface coils in MR technology, relatively good quality images of the brain in the medium-size animals can be obtained using clinical 1.5T MR machines. For the inherent physics of MRI, image quality depends on the objective volume. Thus, the larger the volume of the animal, the better the MR image quality which can be obtained. When fat particles are accidentally exposed to cerebral vessels as in cerebral fat embolism, it evokes reversible brain edema and injury [21]. In previous studies, it is demonstrated that infusion of fat emulsion into the cerebral artery caused reversible disruption of the BBB and brain edema. The reversible change was depicted on MRI studies [8, 20, 22]. In detail, MRI findings of cerebral-fat embolisms include multifocal hyper-intensities in the gray matter and white matter of cerebrum, cerebellum, and brain stem on T2-weighted images in the acute or subacute stage; these lesions are mostly reversible. In experimentally induced cerebral-fat embolism with a single-bolus injection of fat, there are two types of the embolized lesions, with the type showing vasogenic edema in the hyperacute stage reverting to a normal appearance in the subacute stage. It is found that BBB is disrupted within 30 minutes after neutral fat embolization. Hyperintensity on T2-weighted MR images and lesional enhancement on contrast-enhanced T1-weighted MR images are related to defects in the endothelial wall and the early appearance of vasogenic edema. There were minimal histologic findings in the brain tissue [22]. Since this study investigated the effect of neutral fat (triolein) on brain blood vessel system and some comparison studies based on the histopathologic findings revealed that oleic acid is more toxic than neutral fat in the lung and kidney, another study was conducted to evaluate the differences in embolic effects between two groups: those embolized with triolein and those embolized with oleic acid using MR imaging and electron microscopy. In experimental cerebral fat embolism induced by triolein and oleic acid, both vasogenic and cytotoxic edema appeared in the hyperacute stage. While triolein emulsion produced mainly vasogenic edema and resulted in reversible change in the brain, oleic acid produced more severe tissue damage [8, 20].

Based on the mechanical obstruction theory, cerebral fat embolism should be like ischemic stroke. But data demonstrated that cerebral fat embolism is quite different from ischemic infarction in that the lesion shows hyperintensity on T2-weighted images earlier than ischemic infarction and also in parenchymal enhancement. These findings may be caused by the early appearance of vasogenic edema and disruption of the BBB by fat in the hyperacute stage. Clinically, cerebral fat embolism also differs from ischemic infarction in that the disease is self-limiting and silent symptoms of the disease can make early diagnosis difficult. Several MR imaging studies of clinical cerebral fat embolism have been recently published [23-28]. Acute lesions of cerebral fat embolism are demonstrated by hyperintensity on T2-weighted images, hypointensity on T1-weighted images throughout the white and gray matter in the cerebrum, cerebellum, and brain stem. Such lesions are resolved leaving focal lesions in the gray and white matter because of infarction or demyelination [23-25]. Experimental studies into cerebral fat embolism have revealed that the embolized lesions are demonstrated as 2 types of edema, vasogenic and cytotoxic, in the hyperacute stage: type 1 lesions contain both vasogenic and cytotoxic edema, whereas type 2 lesions are due to vasogenic edema only. Reversibility of the embolized lesions in clinical cerebral fat embolism may depend on the type of the lesion, as demonstrated in a previous report [18] and can be presumed at the hyperacute stage.

Even though BBB modulation was shown in fat embolism model using MRI studies, the pathophysiological characteristics at the cellular level was not intensively studied yet. Therefore, another studies were intended to investigate the cellular mechanism of altered BBB permeability and accompanying brain edema [29, 30]. In addition to MR imaging studies, oleic acid induced BBB alteration was visualized by microscopic observation of extravasated Evans blue or FITC-labeled dextran in the brain sections. Dysfunction of BBB was quantitated by measuring Evans blue concentration in the brain tissue or counting the immunoreactivity of endothelial barrier antigen (EBA), a marker protein for endothelial cells displaying barrier characteristics and is only expressed in blood vessels with fully developed BBB properties in the brain. Injection of oleic acid into the carotid artery caused vasogenic edema on MR images and leakage of dextran on FITC-labeled dextran images, presenting as increased vascular permeability. The immunoreactivity of the EBA was decreased at the same brain area. The embolized hemisphere revealed contrast enhancement on the contrast-enhanced image, mild hyperintensity on the T2WI, minimal hyperintensity on the DWI, and isointensity on ADC map image. No leakage was seen in the normal contralateral cortex while FITC-labeled dextran leakage was observed only in the lesion hemisphere [29]. Another study which was designed to investigate the effects of hypothermia in the model of oleic acid induced brain fat embolism also demonstrated that abnormal EBA expression pattern in the endothelial cells, indicating BBB disruption, and reduced number of EBA-positive vessels in the cortex of the lesion hemisphere compared to the normal hemisphere. Brain edema at the tissue level can be measured by weighing the water content of the isolated brain. 6 hours after oleic acid injection, water content in the lesion hemisphere increased and returned to normal range at 24 hours after. This data is similar to MR imaging data indicating reversible brain edema in cerebral fat embolism model [30].

BBB's essential role protecting the brain from the exogenous deleterious substances such as bacteria or toxins in the pathological conditions is critical to keep the brain in a healthy state. But this tight restriction of BBB poses a significant obstacle to the delivery of many drugs into the brain. As a result, drug therapy has rarely been available to treat brain tumors, cancer metastases to brain and other brain diseases and therefore overcoming BBB is critical to the CNS drug developers and significant effort is being made to develop tools which either disrupt or increase the permeability of the BBB [31-33]. Efficient delivery of diagnostic agents across the BBB is also beneficial to improve the detection of pathologic targets inside of brain. Even though BBB is a very restrictive barrier, the phenotype of BBB is disturbed under some pathological conditions. Stroke and traumatic brain injury are accompanied by severe BBB dysfunction. However, pathological dysfunction of the BBB is not appropriate for medical use. Therefore, development of strategies which causes controllable and transient disruption of the BBB will be beneficial for basic research and clinical application. Transient BBB disrupting effect by fatty acid can provide the breakthrough to the drug delivery system across BBB in the future.

Inflammation and Free Radicals

During the investigation of the mechanism of oleic acid induced BBB disruption, it is first expected that cerebral fat embolism model is similar to ischemic brain injury since vascular obstruction by micro fat emboli was considered as the primary pathogenic factor. But experimental data suggested that BBB disruption cannot be explained solely by mechanical obstruction of the brain vessels. The embolized lesions showed no significant difference in cerebral hemodynamic parameters compared with those of the contralateral normal side [34] and non fat micro particle model showed different findings compared with fatty acid model [22]. Chan and Fishman [35] proposed free radical mediated injury theory but this idea was not investigated carefully before. Based on Chan's idea and some preliminary data, hypothesis was made as follows. 1) If free radical is involved in the pathogenesis, there will be cells responsible for inflammation and free radical formation. 2) Evidence of free radical activity should be observed in the vicinity of BBB. 3) Free radical inhibitor should inhibit BBB disruption.

Even though inflammatory substances and free radicals can be produced by many kinds of cells in brain, inflammatory cell is the first candidate. Using cell type specific markers, the presence of inflammatory cell was evaluated. Antibody against myeloperoxidase (MPO), the marker of neutrophil, revealed infiltrated cells in the perivascular and parenchymal area. Activated microglia or macrophage was not observed. Since it is known that neutrophil adheres to the endothelial surface via intercellular adhesion molecule-1 (ICAM-1) before it infiltrates into brain parenchyma, the presence of ICAM-1. With the immunohistochemistry, induction of ICAM-1 in the brain vessels following oleic acid treatment was evaluated. Area with ICAM-1 induction was overlapped with the area of neutrophil infiltration as expected [30]. Even though the number of neutrophil was higher at 24 h, ICAM-1 induction was transient. The short lasting ICAM-1 induction implies that inflammation by oleic acid is a transient phenomenon. Regarding the changes of blood vessels, there was one group who demonstrated oleic acid induced changes of neutrophil itself [36]. They showed that exposure of isolated human neutrophils to oleic acid caused a rapid increase in the cell surface expression and affinity state of CD11b. Oleic acid triggered neutrophil aggregation and neutrophil adherence to confluent cultures of human umbilical vein endothelial cells. Based on the data here and Mastrangelo's data, it can be concluded that fatty acid makes changes both in neutrophils and vascular endothelial cells. Oleic acid makes neutrophils prone to adhere to the endothelium and cause harmful effect on the BBB. It is well known that infiltrating neutrophils are the exogenous source of deleterious molecules to the brain. Within injured brain tissue, neutrophils contribute to the secondary injury by releasing free radicals, generating lipid mediators and activating thrombosis cascades [37, 38]. For the next step, it was investigate whether there is any evidence of free radical activity after fatty acid treatment. Induction of iNOS and presence of nitrotyrosine were observed in the affected brain. In particular, vessels and surrounding areas showed strong immunoreactivity for nitrotyrosine. From these data, it is speculated that neutrophils produce massive amount of nitric oxide and nitric oxide exerts its influence mostly on the vessels. Even though the pathological processes are different from fat embolism model, there are many studies showing the effect of nitric oxide on BBB disruption. Many cerebrovascular diseases have the inflammatory components, and the free radicals including nitric oxide may play an important role in alterations of the BBB observed during disease states [39, 40]. The pathological roles of superoxide and nitric oxide are intensified because nitric oxide reacts with superoxide to form the extremely reactive hydroxyl and peroxynitrite anions, which are decomposed into nitrogen dioxide and hydroxyl radicals. Peroxynitrite damages cells by lipid peroxidation [41] and tyrosine nitrosylation [42]. Nitrotyrosine is specifically derived from the reaction of peroxynitrite with proteins and regarded as a specific marker of peroxynitrite generation [42]. Viswanathan and colleagues [43] suggested that linoleic acid exposure resulted in increased nitrotyrosine levels and the proinflammatory effects of linoleic acid can be mediated through peroxynitrite signaling. In addition to nitrotyrosine, MDA is also produced in our model. MDA is one of the major byproducts of lipid peroxidation and can form covalent adduct with proteins [44]. Covalent binding of MDA with proteins may cause impairment of cellular function and integrity. It is known that lipid peroxidation results from the attack of reactive free radicals on the polyunsaturated fatty acids of the membrane. In spite MDA production was observed in the brain, increase of MDA was not observed in the vessels. Therefore it is expected that MDA may not be the major contributor in BBB disruption. Even though nitric oxide itself is detrimental enough to cause BBB damage, it is expected that there would be other factors contributing to BBB disruption. Studies on mediators of BBB opening have been performed in both *in vivo* and *in vitro* experiments and various factors were suggested [16, 45, 46].

Since inflammation exert its deleterious actions by triggering the induction and activation of MMPs [47, 48], MMP induction was investigated. MMPs are a gene family of neutral proteases important in normal and abnormal brain function. They have been implicated in processes such as (a) opening of BBB by the lysis of basement membrane and extracellular matrix [49-51], (b) invasion of neural tissue by blood-derived immune cells, (c) shedding of cytokines and cytokine receptors, and (d) direct cellular damage in CNS diseases. MMPs can be divided into five main groups based on protein structure: gelatinases (MMP-2 and -9), collagenases (MMP-1, -8, -13, and -18), stromelysins (MMP-3, -10, and -11), membrane-type MMPs (MMP-14, -15, -16, -17, -24, and -25), and matrilysins (MMP-7 and -26). It is reported that neutrophil infiltration contributes to enhanced MMP-9 action in the ischemic brain by releasing MMP-9, which might participate in the inflammatory reaction [52]. Other studies have also shown that MMPs are important for BBB breakdown and edema formation [53, 54]. BBB comprises endothelial tight junctions, pericytes, astrocytic end feet, and basal lamina. The basal lamina contains extracellular matrix (ECM) molecules such as type IV collagen, laminin, and fibronectin and most of which are substrates for MMPs, especially MMP-2 and MMP-9. In this study, MMP-2, -3, -9 and -13 was shown in the vessels parenchymal cells [30]. From these data, it is speculated that oleic acid induced MMP activation is one of the causes of the BBB opening. In summary, Oleic acid induced inflame-mation and free radical formation in the cerebral vessels and surrounding brain tissue was observed including induction of ICAM-1 in the vascular endothelial cells, infiltration of neutrophils into the brain parenchyma, induction of free radicals. Oleic acid also activated MMPs in the blood vessels brain parenchyma.

BBB and Inflammation

Several pathological conditions of the brain such as stroke, neurodegenerative diseases, traumatic injury are associated with inflammatory process and structural and functional abnormalities in the cerebral microvasculature or BBB are frequently observed. Even though attention has been paid to the inflammatory activities of astrocytes and microglia during CNS pathogenesis, the role of the endothelial cells in the brain vasculature and blood derived inflammatory cells is very important as well [55]. Enhanced expression of adhesion molecules on the brain endothelial cells *in vivo* during a pathological state facilitates the entry of leukocytes into the cerebral tissue.

The possibility that the endothelial cells may secrete inflammatory mediators may indicate an important gate function for the blood inflammatory cells between the immune system and the brain parenchyma.

In that respect, the brain endothelial cells may and will contribute both to the onset and to the progression of inflammation of many CNS diseases. As a major component of the BBB which is residing at the interface between the blood circulation and the brain parenchyma, the vascular endothelium of the CNS is positioned to be a critical regulator of CNS environment and function.

Therefore, disturbances to endothelium might exacerbate or trigger neuropathologic conditions. The endothelium of the CNS is extremely dynamic, and is either directly or indirectly interactive with the entire repertoire of brain cells. Altered functioning of endothelium can potentially provoke a cascade of cytopathic activity that leads to extensive death and destruction throughout the CNS parenchyma. It is reported that perturbations of the CNS endothelium are closely linked to the pathophysiology of several inflammatory, infectious, and degenerative disease states of the CNS.

The endothelium is an extraordinarily active interface between the blood and the tissues. It has a definitive role in the normal physiology of blood distribution and coagulation and a critical role in inflammation and injury. Under healthy conditions, the BBB not only regulates the entry of endogenous or exogenous compounds into the brain, but also restricts the cellular infiltration into the brain as well. The normal endothelial cell layer provides vascular inner surface that prevents platelet and leukocyte adhesion and activation of coagulation system. The highly specialized CNS endothelium forms a tight barrier which isolates the brain from peripheral immune system, and allows only a few inflammatory cells to migrate into the CNS. The low expression of major histocompatibility complex (MHC) antigens, the low number of antigen-presenting cells in the healthy CNS, and the fact that the CNS is not properly drained by a fully developed lymphatic vasculature, make the brain an "immunosecluded" site [56, 57]. However, when inflammation occurs, an extensive leukocyte migration into the brain takes place, for instance during cerebral ischemia, multiple sclerosis, Alzheimer's disease, or brain infections.

Besides the CNS endothelial cells and neutrophils, other cell types such as microglial cells and perivascular macrophages may eventually be involved in the neuroimmune response. The barrier function of the BBB can change dramatically during various diseases of the CNS.

Enhanced BBB permeability is considered to be the result of either opening of tight junctions or of enhanced pinocytotic activity and the formation of transendothelial channels. The BBB itself may play an active role in the mediation of the neuroimmune response either by production of inflammatory mediators or by the expression of adhesion molecules. Disruption of the BBB function and immune response may be related to the factors such as chemokines, cytokines, and other related factors.

Cerebral Hemodynamics

Cerebral fat embolism in human patients shows reversible and/or irreversible change in MR images and brain pathology [58, 59]. Experimental cerebral fat embolism by a bolus injection of fat emulsion also shows similar findings in animal models [8, 60]. Experimental cerebral fat embolism induced by small particles of neutral fat has revealed major findings of reversible change.

The MRI findings of the embolized lesions were prominent at 1 hour, less evident on day 1, and reverted to normal after day 4 [61]. Based on the 2 major theories of pathogenesis of fat embolism, so called mechanical and chemical theories, we expect that there may be hemodynamic changes in the brain by fat embolism.

With the view of mechanical theory supporters, fat embolism depends upon many factors, such as the size and shape of the fat particles, the hydrostatic pressure in the vascular bed, and the flow characteristics of the vessels into which the fat particles have been introduced. The size of the fat particles may be one of the most important factors in the pathogenesis of fat embolism.

From the point of chemical theory proponents, fat embolism also can interfere with the hemodynamics. Endothelial cells are very sensitive to various stimuli including inflammatory mediators and produce and release all kinds of vasoactive substances into the blood stream. These vasoactive substances can modulate vessel wall activities and then finally blood circulation will be changed as a result.

Since the changes in the MRI appeared predominantly during the hyperacute stage after embolization and embolized lesion showed reversibility, it is speculated that cerebral hemodynamic change in the lesion will be changed in the hyperacute stage. One study was aimed to evaluate relative cerebral blood volume (rCBV), cerebral blood flow (rCBF), and mean transit time (MTT) in the hyperacute stage of cerebral fat embolism induced by triolein emulsion [34].

In this study on perfusion-weighted images (PWIs), there were no significant differences in the rCBVs, rCBF, and MTT in the lesions or in the contralateral normal side at each acquisition time. The rCBV, rCBF, and MTT ratios of 2, 4, and 6 hours showed no significant change compared with those at 30 minutes. In the hemodynamic study with PWI, the mean values of rCBV, rCBF, and MTT of the lesion showed no statistical difference compared with those of the contralateral side at each acquisition time. There were no any significant changes in the ratios of rCBV, rCBF, and MTT over time as well. The authors concluded that the embolized lesions induced by fat emulsion showed no significant difference in cerebral hemodynamic parameters from those on the contralateral normal side and the result suggest that consideration of the hemodynamic factor of embolized lesions is not necessary in further studies of the BBB with fat emulsion [34]. It is not clear why the brain hemodynamics was not significantly modified by fat embolism in this study but if monitoring of the blood flow with more sensitive systems is possible, there will be more sophisticated data to explain the brain hemodynamics in the future.

Conclusion

Inflammation is one of the contributing factors of oleic acid induced BBB disruption as summarized in figure 1. By stimulating endothelial cells and neutrophils, a cascade of inflammation is induced by oleic acid in the brain microvasculature. Inflammatory reaction propagated and expanded by free radical formation and MMP activation.

By degrading components of BBB structure, functional and structural integrity of BBB is disrupted. But this inflammatory cascade is not severe enough to cause damage in neighboring cells such as neurons or astrocytes. Oleic acid induced damage is limited by its duration and range and therefore BBB disruption and brain edema is transient during the hyperacute phases.

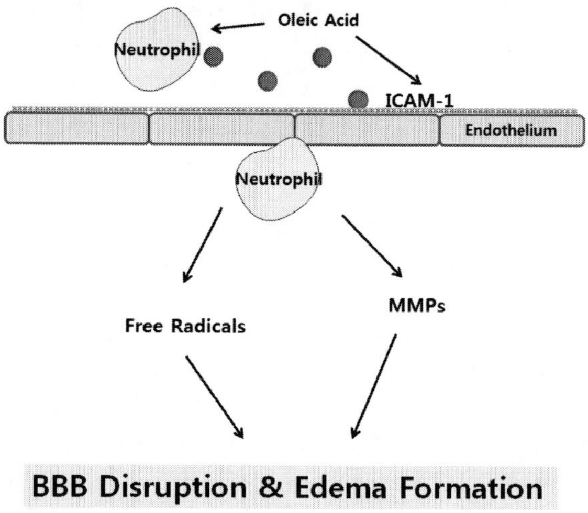

Figure 1. Pathogenesis of oleic acid induced BBB dysfunction. Oleic acid stimulates endothelial cells and neutrophils. Activated endothelial cells produce ICAM-1 and attract neutrophils into the brain. Activated neutrophils trigger inflammatory cascade including free radical formation and MMPs activation which lead to transient BBB disruption and brain edema formation as a final consequence.

REFERENCES

[1] R. W. Ordway, J. V. Walsh Jr. and J. J. Singer, *Science* 244, 1176 (1989).

[2] K. D.Philipson and R. Ward, *J. Biol. Chem.* 260, 9666 (1985).

[3] P. Steneberg, N. Rubins, R. Bartoov-Shifman, M. D. Walker and H. Edlund, *Cell Metab.* 1, 245 (2005).

[4] W. M. Pardridge, J. D. Connor and I. L. Crawford, *CRC Crit. Rev. Toxicol.* 3, 159 (1975).

[5] C. Samieri, C. Féart, C. Proust-Lima, E. Peuchant, C. Tzourio, C. Stapf, C. Berr and P. Barberger-Gateau, *Neurology* 77, 418 (2011).

[6] K. Taviloglu and H. Yanar, *Surg. Today* 37, 5 (2007).

[7] S. Akhtar, Anesthesiol. Clin. 27, 533 (2009).

[8] H. J. Kim, J. H. Lee, C. H. Lee, S. H. Lee, T. Y. Moon, B. M. Cho, H. K. Kim, B. P. Park and K. H. Chang, *Am. J. Neuroradiol.* 23, 1516 (2002).

[9] I. Wilhelm, C. Fazakas and I. A. Krizbai, *Acta Neurobiol. Exp.* 71, 113 (2011).

[10] S. U. Kim and J. de Vellis, J. Neurosci. Res. 81, 302 (2005).

[11] V. H. Perry, Brain Behav. Immun. 18, 407 (2004).

[12] G. J. Guillemin and B. J. Brew, *J. Leukoc. Biol.* 75, 388 (2004).

[13] A. Tsuji, NeuroRx. 2, 54, (2005).

[14] N. J. Abbott and I. A. Romero, Mol. Med. Today 2, 106, (1996).

[15] L. González-Mariscal, A. Betanzos, P. Nava and B. E. Jaramillo, *Prog. Biophys. Mol. Biol.* 81, 1 (2003).

[16] N. J. Abbott, L. Rönnbäck and E. Hansson, *Nat. Rev. Neurosci.* 7, 41 (2006).

[17] G. A. Dhopeshwarkar and J. F. Mead, *Adv. Lipid Res.* 11, 109 (1973).

[18] F. Kamp, H. V. Westerhoff and J. A. Hamilton, *Biochemistry* 32, 11074 (1993).
[19] F. Kamp, D. Zakim, F. L. Zhang, N. Noy and J. A. Hamilton, *Biochemistry* 34, 11928 (1995).
[20] H. J. Kim, Y. S. Pyeun, Y. W. Kim, B. M. Cho, T. H. Lee, T. Y. Moon, K. T. Suh and B. R. Park, I*nvest. Radiol.* 40, 270 (2005).
[21] D. J. Butteriss, D. Mahad, C. Soh, T. Walls, D. Weir and D. Birchall, *Am. J. Neuroradiol.* 27, 620 (2006).
[22] H. J. Kim, C. H. Lee, S. H. Lee, B. M. Cho, H. K. Kim, B. R. Park, S. Y. Ye, G. R. Jeon and K. H. Chang, *Invest. Radiol.* 36, 460 (2001).
[23] H. Chrysikopoulos, V. Maniatis, J. Pappas, P. Filalithis, C. Gogali and Sfyras, *Clin. Radiol.* 51, 728 (1996).
[24] A. Saito, K. Meguro, A. Matsumura, Y. Komatsu and N. Oohashi N, *Neurosurgery* 26, 882 (1990).
[25] M. E. Finlay and M. D. Benson, *Clin. Radiol.* 51, 445 (1996).
[26] K. Wegener, F. Bolgert and C. Pierrot-Deseilligny, *Eur. Neurol.* 42, 65 (1999).
[27] A. Yoshida, Y. Okada, Y. Nagata, K. Hanaguri and M. Morio, *J. Trauma* 40, 437 (1990).
[28] A. di Summa, A. Beltramello, G. B. Fratucello, L. G. Bongiovanni, G. Zanette and A. Polo, *Eur. Neurol.* 40, 55 (1998).
[29] J. W. Park, H. J. Kim, G. S. Song and H. S. Han, *J. Korean Neurosurg. Soc.* 47, 203 (2010).
[30] K. M. Lee, J. H. Jang, J. S. Park, D. S. Kim and H. S. Han, *Genes and Genomics* 31, 89 (2009).
[31] R. Cavaliere and D. Schiff, *Neurosurg. Focus* 22, E6 (2007).
[32] J. Drappatz and P. Y. Wen, Expert Rev. Neurother. 6, 1465 (2006).
[33] M. A. Bellavance, M. Blanchette and D. Fortin, AAPS J. 10, 166 (2008).
[34] Y. W. Kim, H. J. Kim, B. M. Cho, T. Y. Moon and C. K. *Eun, Am. J. Neuroradiol.* 27, 398 (2006).
[35] P. H. Chan and R. A. Fishman, *J. Neurochem.* 35, 1004 (1980).
[36] A. M. Mastrangelo, T. M. Jeitner and J. W. Eaton, *J. Immunol.* 161, 4268 (1998).
[37] F. C. Barone, *J. Cereb. Blood. Flow. Metab.* 19, 819 (1999).
[38] G. C. del Zoppo and J. M. Hallenbeck, *Thromb. Res.* 98, 73 (2001).
[39] J. E. Merrill and S. P. Murphy, *Brain Behav. Immun.* 11, 245 (1997).
[40] R. D. Hurst and M. S. Dobbie, *Biochemist* 23, 15 (2001).
[41] H. Ischiropouls, T. Beers, T. Ohnishi, D. Fisher, S. E. Garner and S. R. Tom, *J. Clin. Invest.* 97, 2260 (1996).
[42] A. Gow, D. Duran, S. R. Thom and H. Ischiropoulos, *Arch. Biochem. Biophys.* 333, 42 (1996).
[43] S. Viswanathan, W. Guoyao, T. Michal and H. Bernhard, *J. Lipid Res.* 45, 794 (2004).
[44] H. Mahmoodi, M. Hadley, Y. X. Chnag and H. H. Draper, *Lipids* 30, 963 (1995).
[45] N. J. Abbott, *J. Anat.* 200, 629 (2002).
[46] G. M. William, *Microcirculation* 8, 89 (2001).
[47] Y. Gursoy-Ozdemir, A. Can and T. Dalkara, *Stroke* 35, 1449 (2004).
[48] J. H. Heo, S. W. Han and S. K. Lee, *Free Radic. Biol. Med.* 39, 51 (2005).
[49] M. Asahi, X. Wang, J. Jung, T. Mori, T. Sumii, M. A. Moskowitz, M. E. Fini and E. H. Lo, *J. Neurosci.* 21, 7724 (2001).

[50] A. Lukes, S. Munn-Bryce, M. Lukes and G. A. Rosenberg, *Mol. Neurobiol.* 19, 267 (1999).

[51] S. Munn-Bryce and G. A. Rosenberg, *J. Cerb. Blood Flow Metab.* 18, 1163 (1998).

[52] C. Justicia, J. Panes, A. Cervera, R. Deulofeu, A. Chamorro and A. M. Planas, *J. Cerb. Blood. Flow. Metab.* 23, 1430 (2003).

[53] B. T. Hawkins, T. F. Lundeen, K. M. Norwood, H. J. Brooks and R. D. Egleton, *Diabetologia* 50, 202 (2007).

[54] M. A. Kelly, A. Shuaib and K. G. Todd, *Exp. Neurol.* 200, 38 (2006).

[55] H. S. Han and K. Suk, Neurovasc. Res. 2, 409 (2005).

[56] D. A. Hafler and H. L. Weiner, *Immunol. Rev.* 100, 307 (1987).

[57] H. Wekerle, C. Lington, H. Lassmann and R. Meyermann, *Trends Neurol. Sci.* 9, 271 (1986).

[58] S. Sevitt, *Ann. Clin. Res.* 9, 173 (1977).

[59] M. Takahashi, R. Suzuki, Y. Osakabe, J. I. Asai, T. Miyo, G. Nagashima, T. Fujimoto and Y. Takahashi, *J. Trauma* 46, 324 (1999).

[60] H. J. Kim, C. H. Lee, S. H. Lee and T. Y. Moon, *Invest. Radiol.* 38, 625 (2003).

[61] H. J. Kim, C. H. Lee, H. G. Kim, S. D. Lee, S. M. Son, Y. W. Kim, S. M. Kim and C. K. Eun, *Am. J. Neuroradiol.* 25, 958 (2004).

In: Oleic Acid: Dietary Sources, Functions and Health Benefits ISBN: 978-1-62618-332-2
Editor: Luciano Paulino Silva

Chapter 9

FOOD SUPPLEMENTATION WITH VEGETABLE OILS RICH IN OLEIC ACID: HEALTH-PROMOTING EFFECTS AND APPLICATION VIABILITY

Fabiano J. Contesini*, Elaine B. Ceresino, Ruann J. S. Castro, Francisco F. C. Barros, Evelyn Sassaki and Hélia H. Sato

Department of Food Science, College of Food Engineering, State University of Campinas (UNICAMP), Campinas, SP, Brazil

ABSTRACT

The role of food in human health is broad due to its diverse functions. Many foods may be explored in different aspects, since they are not only important for basic nutrition, but also show interesting functional properties. Unsaturated oils are commonly used in food supplementation, taking into account that they are sources of fatty acids, such as oleic acid. The regular consumption of these foods is associated with healthy effects on the human body, including reduction of coronary heart disease risk and beneficial effects on blood lipids, including reduction of total and LDL-cholesterol, triglycerides and increase of HDL-cholesterol in blood. The substitution of saturated for unsaturated fatty acids, as well as the supplementation with vegetable oils rich in oleic acid have been used in food technology as a strategy of food production, with improvement in functional and sensory characteristics. Milk is an example of these substitutions, widely applied in the food industry. Various research studies have shown that the consumption of milk enriched with oleic acid promoted a myriad of positive effects in healthy volunteers, subjects with increased risk factors and cardiovascular patients. The effects of supplementation with oleic acid in different types of salami and other products of animal origin have also been reported. Wheat breads supplemented with oleic-rich sunflower seed were produced without any significant adverse effects regarding the crust color, crumb grain structure and uniformity. Oleic acid has also been used in edible coatings such as plasticizer and emulsifier, for example, to preserve the quality of cold-stored

* E-mail address: fabiano.contesini@gmail.com.

strawberries, carrots and other fresh products. They improve sample appearance and confer significant property barriers. These are important sensory and technological characteristics that justify the process viability, besides the addition of high nutritional value. Therefore, food supplementation with vegetable oils rich in oleic acid can be very interesting, and different studies about this topic are going to be reviewed in this manuscript.

INTRODUCTION

The oleic acid (18:1) is a monounsaturated omega-9 fatty acid, commonly found in both animal and plant lipids. It can be used as an ingredient for soap making, a surfactant agent and an excipient in pharmaceuticals, among many other uses.

It is known that products like soybean oil and olive oil are rich sources of unsaturated fatty acids. The oleic acid is among them and represents one of the most studied fatty acids because it shows many benefits when applied as a food supplement and enrichment element.

The main difference between food enrichment and food supplement lies in the idea that enrichment is the restoration of nutrients that were lost or removed during processing; whereas supplemenation is the addition of nutrients, especially through a mixture of foods. Furthermore, fortification is the addition of nutritional value when it is used to match another parallel product: an example is the addition of vitamins to margarine, so that it will be equivalent to butter [1].

Results of many studies also showed positive effects of ingestion of milk enriched with oleic acid on risk factors for cardiovascular diseases [2-4]. Moreover, wholegrain and refined wheat breads were supplemented with high-oleic sunflower seed, generating sensory acceptance and higher nutritional value [5].

Regarding habitual elaidic acid- and palmitic acid-rich diets, the oleic acid-rich diet significantly lowered LDL cholesterol and plasma cholesterol concentrations, whereas HDL cholesterol levels were not significantly affected [6].

A high amount of monounsaturated fatty acids consumed in a diet can reduce levels of insulin requirement in the body, prevent cardiovascular diseases and decrease plasma concentration of glucose. As such, a healthy diet must limit the amount of saturated fatty acids, reducing the total cholesterol level [7].

In an attempt to replace saturated fats and hydrogenated oils, high-oleic oils are being developed to provide higher stability in cooking oils. This will not only deliver substantial public health benefits, but also efficiently replace oils with high contents of palmitic acid and *trans* fatty acids [8].

This chapter reviews the use of edible oils rich in oleic acid, their processing and the utilization of oleic acid as a supplement for food, including several patents with this application. The health effects of oleic acid are also reported, emphasizing the importance of the substitution of unsaturated fats and oils in the place of saturated fats and oils.

PROCESSING OF EDIBLE OILS RICH IN OLEIC ACID

Ingestion of lipids, which have been subject to different processing, has been studied to evaluate the consequences in human and animal organisms. Thermal stress and oxidation bring up questions that might be discussed about nutritive value and safety.

During frying, fats are in contact with oxygen, water and pro-oxidants, which, when associated with high-temperatures, may suffer oxidation, polymerization and hydrolyses [9, 10]. The oils and fats used for commercial frying applications must be stabilized in order to avoid the formation of these byproducts [9]. The frying process is one of the most common practices used to cook food in both home and industrial operations and is also the essence of the fast-food industry, being that it provides tasty, ready-to-eat food in a relatively short period of time. These are important reasons considering the highly aggressive process involving high temperature, and that is why the stability of the oil is a crucial factor when choosing a frying medium. So, partially hydrogenated oils have been commonly used as the frying media because of their stabilities [10]. However, during the hydrogenation process a high quantity of *trans* fatty acids (TFA) is formed and the use of saturated fats has been avoided due to health implications, as well as TFA. Their consumption has been linked to the increase of serum levels of low-density lipoprotein cholesterol and the decrease of high-density lipoprotein cholesterol, while TFAs are more specifically related to proinflammatory effects [11, 12].

Because of these health issues and the *trans* fats labeling requirements by government organizations of different countries around the world, the food industry has been trying to find alternatives to partially hydrogenated oils. A number of fats and oils used in frying and other high-temperature applications have been carefully studied, and altering fatty acid compositions of oilseeds by plant breeding is now commonly done to improve the oxidative stability of oils [10, 13].

Gerde et al. used no-*trans*, low-linolenic acid soybean oils and tested their frying performance [10]. Typical soybean oil with about 7–8% linolenic acid (18:3) is known to be oxidatively unstable, especially during frying. They observed a reduction of linolenic acid and an increase of linoleic and oleic acid levels on the modified oil when compared to control oil, and these values were used in the calculation of oxidizability. They concluded that ultra low-linolenic acid oil had the highest theoretical stability, followed by low-linolenic acid, and the control had the highest value of oxidizability, as expected. High oleic acid soybean oil (HOSBO) and low linolenic acid soybean oil (LLSBO) were evaluated to determine frying oil stabilities and the flavor quality and stability of potato chips by Warner and Gupta [13]. The HOSBO has 85.1% of oleic acid, while LLSBO has 26.1%. Although HOSBO had an excellent oxidative stability during frying of potato chips, this product presented some flavor problems that could not be observed by instrumental analysis. On the other hand, LLSBO had the lowest frying stability, compared to HOSBO, their blends and control (cottonseed oil), but sensory data indicated that potato chips fried in LLSBO showed better overall quality scores than those fried in HOSBO.

Besides the fact that frying is a high-temperature and consequently very aggressive treatment, roasting may also affect lipid oxidation. To contour this problem, high-oleic peanuts were developed, and the effect of roasting was evaluated by O'Keefe et al. with the intention of preventing off-flavors in peanut-derived products [14]. The shelf life of roasted high-oleic peanuts is greater than that of conventional roasted peanuts, and Bolton and Sanders also concluded that the roasting oil with a high ratio of oleic/linoleic acid increases the shelf life of roasted peanuts [15].

Another widely applied process to vegetable oils is hydrogenation – for sure the most versatile between all the treatments. The aim of this process, when applied to edible oils, is adapting their melting point according to industry necessities and improving their stability, which, as cited before, is very important to prevent further deteriorating reactions [16]. The hydrogenation process consists basically of the direct addition of hydrogen to the double bonds in the fatty acid chains of the triglycerides, using nickel on silica or alumina as catalysts due to their availability, low cost and inert nature. This reaction occurs in a reaction vessel with controlled temperature and pressure, where gaseous hydrogen, oil and nickel catalysts are placed [17, 18]. In addition to saturation of double bonds, *cis-trans* isomerization also occurs, which leads to a formation of diverse products [19]. During the process, linolenic ester is hydrogenated to linolenic acid (18:2), which is hydrogenated to oleic acid or oleic *trans* isomer – elaidic acid. The rates of these reactions are different, and the final reaction products directly influence the sensory characteristics of the product developed and certain health effects. Brat and Pokorný demonstrated that in margarines and cooking fats available in the Czech market, the main fatty acids presented on their composition were oleic and linoleic acids [20]. They also reported negligible *trans* fatty acid contents. It is very well established that the intake of saturated fat, *trans* fat and cholesterol should be as low as possible as part of a nutritionally balanced diet. These substances are linked with the raising of blood LDL cholesterol levels, which increase risk of coronary heart disease, which is greatest cause of death in the US.

TECHNOLOGICAL APPLICATIONS OF OLEIC ACID

The application of oleic acid as a supplement in food is a very old practice. The use of vegetable oils has existed since the ancient ages with technical objectives such as improving texture, flavor and food preservation, for example. The development of food technology in the second half of the twentieth century has not changed that paradigm. The improvements have only sophisticated and developed these processes with the same previous goals. Only recently, with the advancement of knowledge of the effect of fatty acids on human health, a new approach was given to this supplementation. At present, we know much about the benefits of some types of fatty acids on health. Obviously, this resulted in a great increase of interest in products that provide these compounds and their health benefits.

The commercial interest in the application of oleic acid as a supplement or replacement of all or part of lipids in foods is evidenced by the large number of patents related to this field (Table 1). Likewise, it is possible to detect the emergence and implementation of this technology. Strong growth was achieved in the eighties, when, especially in Japan and Western Countries, a large number of patents began to be deposited. These patents range from replacement and/or oleic acid supplementation in infant food, frozen foods, dairy drinks, and others. From the year 1990, a new trend was observed: it was the beginning of a new generation of products in which new processing technology would aggregate methods that had been previously applied. All of this behavior was reinforced by a greater exposure to consumer marketing campaigns and knowledge related to human health.

A simple example of the change in this kind of technology can be found in patent EP 012990, which relates an edible fat-containing food product adapted for human infant nutrition containing a fat composition [21]. Basically, the invention describes a fat composition that comprises palm oil; an oil selected from olive oil and oleic oil; an oil selected from coconut oil, babassu oil and palm kernel oil; an oil selected from soybean oil, corn oil, sunflower seed oil, cotton seed oil and safflower oil; and, if desired, up to 2% calculated on the weight of the fat composition, of a lecithin, such as soy lecithin. The amounts of the oils are so chosen that the composition contains defined quantities of linoleic acid, oleic acid, lauric and myristic acids and palmitic and stearic acids. This technology was cited in patent EP0484266, which describes a fat mixture for infant and adult nutrition, basically comprised of a mixture of phospholipids from the cerebrum of domestic animals, at least one vegetable oil, and/or at least one animal fat, and/or at least one fish oil, and/or medium chain triglycerides in which the ratios between oleic/linoleic/alpha-linolenic fatty acids, and between arachidonic/docosahexaenoic fatty acids, and the phospholipid content are similar to those of human milk and the Mediterranean diet [22]. Until September 2011, this patent was cited in 32 other patent documents.

Table 1. Patents in Epodoc base with the terms *Oleic and Acid(s) and Food(s)*

Year/period	Results of search[1]
Before 1969	21
1970-1979	12
1980-1989	31
1990-1999	70
2000-2009	155
2010-2011	22

Note: [1] Epodoc patent base 12/09/2011.

Another example is found in document US 6136858 [23]. The invention is directed toward an improved infant formula containing a lipid blend that softens the firmer stools associated with typical infant formula. A specific formula in accordance with the invention is comprised of carbohydrates, proteins, vitamins and minerals and a lipid mixture of high oleic safflower oil, soy oil and coconut oil at specific levels and ratios. The invention also describes mixtures of fatty acids that provide infant stool patterns more closely resembling that of a breast-fed infant. Another patent document is EP 0496456, which relates to substitute milk fat compositions for replacing at least part of the milk fat necessary to feed young mammals and especially infants [24]. The invention also relates to their preparation by rearrangement of fats using lipase enzymes as rearrangement catalysts, and to milk substitutes containing fat substitution or supplementation with oleic acid, health effects and intervention studies.

SUPPLEMENTATION OF MILK WITH OLEIC ACID: HEALTH EFFECTS AND INTERVENTION STUDIES

In the last years, an extraordinary increase in the number of functional foods available has been observed. Several studies have shown that modification of fat consumption, mainly free fatty acids, can improve the lipid and carbohydrate metabolism, promoting the decrease of cardiovascular disease risk. In fact, a lower coronary heart disease incidence and total mortality have been associated with the inclusion of foods that are a source of unsaturated fatty acids, for example, fish and olive oil. Some properties such as antiarrhythmic, hypolipemic, antithrombotic and anti-inflammatory properties have been suggested to explain the cardioprotective effect of these fatty acids [4].

The substitution of saturated fatty acids by unsaturated fatty acids, as well as the supplementation with vegetable oils rich in oleic acid has been used in food technology as a strategy of food production, with improvement in functional and sensory characteristics. Milk is considered a ubiquitous food and in the traditional diet has varied greatly in different regions of the world, since humans continue to consume it mainly beyond infancy. According Lopez-Huertas "milk and dairy products are every day consumed foods and constitute a good and popular source of bioavailable calcium [25]. Whilst milk and dairy foods in general are perceived as healthy foods, milk fat contains about 70% of saturated fatty acids. These fatty acids (mainly myristic and palmitic acids) may raise total and LDL cholesterol, increasing the risk of CV diseases; hence the consumption of milk fat should be limited. Whilst this substitution should produce nutritional benefits derived from the overall reduced intake of saturated fat, many of the final products obtained claim beneficial effects on blood lipids, i.e. reductions on total and/or LDL-cholesterol, triglycerides or increases of HDL-cholesterol in blood. Baró et al. studied the effect of a commercially available skimmed milk supplemented with polyunsaturated fatty acids, oleic acid, and vitamins E, B6 and folic acid on risk factors for cardiovascular disease [2]. The study was conducted for 12 weeks and consisted of an administration of 500mL/day of semi-skimmed milk for 4 weeks and then 500mL/day of the supplemented milk for 8 more weeks. As a consequence, a decrease in plasma concentration of total and LDL cholesterol was observed. According to the authors, the high level, in enriched milk, of oleic acid (almost 3 times the amount of oleic acid occurring in standard milk) in combination with polyunsaturated fatty acids were responsible for the reduction of LDL and total cholesterol.

Carrero et al. in a similar study, evaluated the consumption of milk enriched with ω-3 polyunsaturated fatty acids, oleic acid, vitamins E and B6 and folic acid on risk factors for cardiovascular disease in thirty subjects ages 45 to 65 years with mild hyperlipidemia [3]. It showed that consumption of milk supplemented with ω-3 polyunsaturated fatty acids, oleic acid, and vitamins may be useful for decreasing risk factors for cardiovascular disease, since the levels of triacylglycerol, total cholesterol and LDL cholesterol decreased significantly in the group that consumed this milk for 2 months. It is important to highlight that this study was conducted in the context of a Mediterranean lifestyle in which the consumption of some fatty acids is high, for example the oleic acid. According Carrero et al. the lowering effects of oleic acid on total and LDL cholesterol levels have been described extensively and are associated with a wide range of physiologic effects [3]. If this study had been conducted in populations with other dietary patterns, administration of this supplemented milk may produce a greater influence at this level.

Benito et al. evaluated the administration of daily doses of 500 mL of milk enriched with ω-3 and oleic fatty acids, folic acid and vitamin E in patients with metabolic syndrome, which naturally have an increased risk of cardiovascular disease [4]. The research involved 72 patients allocated in two groups: the control group consumed 500mL of semi-skimmed milk per day and the test group consumed 500 mL of enriched milk per day for 3 months. A decrease in serum total cholesterol, LDL cholesterol, triglycerol, apolipoprotein B, glucose and homocysteine was observed in the test group. On the other hand, these parameters had no change in the control group.

Romeo et al. examined the effects on CV risk biomarkers of a daily intake of milk enriched with long chain polyunsaturated fatty acids, oleic acid, carbohydrates, vitamins, and minerals and low in saturated fatty acids in 107 healthy children aged 8-14 years for 5 months [26]. Children were assigned to two study groups: I) a supplemented group, who consumed 600 mL/day of an enriched dairy product and II) a control group, who consumed 600 mL/day of standard whole milk. The impact of milk ingestion on total and LDL cholesterol in the two groups was similar: there was decrease of 3.40 and 3.03% in total cholesterol and 3.11 and 3.50% in LDL cholesterol, respectively, in the control and supplemented groups. The results found in this study contrast with other studies in which the supplementation with polyunsaturated fatty acids and oleic acid may reduce cholesterol levels. The authors attributed the differences to childhood physiological development and other age-related differences.

Edible Coats and Films: Oleic Acid as Moisture Barrier

Edible films are a thin layer of material that can be consumed and work as a barrier to the movement of water vapor, oxygen and solutes to the food [27]. There is a great interest in production and utilization of these coatings, being that they increase quality, stability and shelf life and are used to sustain product freshness and inhibit microbial spoilage. They can be prepared from various biopolymers such as carbohydrates, proteins, lipids and the combination of such [28]. The selection of composition of edible films is very important because their constitution affects their function properties, which have to agree with the purpose and characteristics of the product that will be coated. Proteins and carbohydrates are widely used, and they provide excellent sensorial and mechanical properties, and work as effective barriers for aroma compounds and light gases, such as oxygen and carbon dioxide [29]. However, these compounds present a highly hydrophilic nature that limits their ability to provide a significant moisture barrier.

Oleic acid is incorporated extensively in films and coatings, being that it confers important property barriers. To prevent food quality deterioration, it is important to prevent migration of diverse compounds such as oxygen, volatile flavors and mainly water [30]. One way of improving moisture barrier properties is to add food-grade lipids, which include oleic acid or waxes. The emulsified edible film increases the barrier against water vapor, protecting the product from moisture or desiccation, according to the characteristics of the food matrix.

Many papers report the use of oleic acid as plasticizer or water vapor barrier in films. Despite the relative polarity of oleic acid and its surfactant characteristics, it was confirmed that oleic acid decreases the permeability of protein-based films. Waxes are very effective as moisture barriers, but they have an opaque and matte effect. Many tests were done mixing the two in order to obtain the best characteristics of both and a better performance of high hydrophilic films. For example, oleic acid isolated or mixed with beeswax increased the elasticity, flexibility and stretchability, and reduced the vapor permeability of sodium caseinate-based films [28]. In soy protein isolate (SPI) films, which are also yellowish and poor water vapor barriers, the oleic acid and beeswax were incorporated and the lipids could then be recommended to decrease moisture permeability, increase flexibility and stretchability and maintain gloss. The constant increase in the amount of lipids did not improve the water barrier, but made the films more colorless, so they may be adequate in regards to required sensorial characteristics [29].

CONCLUSION

Triglyceride esters of oleic acid make up a major part of products like olive oil, soybean oil and peanut oil, among others, and they are also present in animal fats. During their processing, thermal stress and oxidation alter the fat composition, giving origin to products that may cause health problems, among them, coronary diseases. With the aim of improving the fat stability, hydrogenation is extensively used, but it may form mainly elaidic acid, a *trans* isomer of acid oleic. The consumption of these *trans* fats, as well as saturated fats and cholesterol should be as low as possible, and many efforts have been made to change this scene.

With the advancement of knowledge related to the importance of fatty acids in human health, an increasing commercial interest of oleic acid as a supplement or replacement of other constitution lipids has been noticed. This is evidenced by a large number of patents that reflect this concern, as well as the extensive quantity of studies that report the utilization of this fatty acid in milk, which is considered a ubiquitous food. Many of the final products obtained claim beneficial effects on blood lipids, i.e. reductions in total and LDL-cholesterol, triglycerides or increases of HDL-cholesterol in blood.

Oleic acid is also an important water vapor barrier and is used in edible films with the intention of increasing the quality, stability and shelf life of fresh products. The population's awareness of a balanced diet is increasing, reflected by the demand for fruits and vegetables with minimal thermal processing. To attend this demand is important to guarantee product freshness and inhibit microbial spoilage. Hence, oleic acid is one of the most important fatty acids to be applied as a supplement or substitute in industrial products with an immense number of advantages.

REFERENCES

[1] J. G. Chopra, *AJPH, 64*, 19, (1974).

[2] L. Baro, J. Fonollá, J. L. Peña, A. Martínez-Férez, A. Lucena, J. Jiménez, J. J. Boza and E. López-Huertas, *Clin. Nutr. 22*, 175 (2003).

[3] J J. Carrero, L. Baró, J. Fonollá, M. González-Santiago, A. Martínez-Férez, R. Castillo, J. Jiménez, J. J. Boza and E. López-Huertas, *Appl. Nutr. Invest. 20*, 521 (2004).

[4] P. Benito, J. Caballero, Moreno, J. C. Gutiérrez-Alcántara, C. Muñoz, G. Rojo, S. Garcia and F. C. Soriguer. *Clin. Nutr. 25*, 581 (2006).

[5] B. Škrbić and B. Filipčev, *Food Chem. 108*, 119 (2008).

[6] P. Nestel, M. Noakes, B. Belling, R. McArthur, P. Clifton, E. Janus and M. Abbey, *J. Lipid Res. 33*, 1029 (1992).

[7] M. D'Imperio, G. Dugo G, M. Alfa, L. Mannina L and A. L. Segre, *Food Chem. 102*, 956 (2007).

[8] Q. Liu, S. Singh and A, Green, *J. Am. Coll. Nutr. 2*, 205S (2002).

[9] K. Warner and S. Knowlton, *JAMA, J. Am. Med. Assoc. 74*, 1317 (1997).

[10] J. Gerde, C. Hardy, W. Fehr and P. J. White, *JAMA J. Am. Oil. Chem. Soc. 84,* 557 (2007).

[11] R. P. Mensink and M.B. Katan, *New Engl. J. Med. 323*, 439 (1990).

[12] D, Mozaffarian, M. B. Katan, A. Ascherio, M. J. Stampfer and W. C. Willett, *N. Engl. J. Med. 354*, 1601 (2006).

[13] K. Warner and M. Gupta. *J. Food Sci. 70*, 395 (2005).

[14] S. F. Okeefe, V. A. Wiley and D. A. Knauft, *J. Am. Oil. Chem. Soc. 70,* 489 (1993).

[15] G. E. Bolton and T.H. Sanders, *J. Am. Oil Chem. Soc. 79*, 129 (2002).

[16] M. J. Mendes, O. A. A. Santos, E. Jordão and A. M. Silva, *Appl. Catal. A-General 217*, 253 (2001).

[17] W. E. Farra and G. R. Listb. Hydrogenation Tecniques. In: N. T. Dunford and H. B. Dunford. *Nutritionally Enhanced Edible Oil Processing*. Routledge, USA, 2004, 320 pp.

[18] E. A. Cepeda and B. Calvo, *J. Food Eng. 89*, 370 (2008).

[19] J.W. Veldsink, B. G. Muuse, M. M. T. Meijer, F. P. Cuperus, R. L. K. M. van de Sande, and K. P. A. M. van Putte, *Fett/Lipid 7*, 244 (1999).

[20] J. Brat and J. Pokorný, (2000) *J. Food Compos. Anal. 13*, 337, (2000).

[21] Arthur Walter Thomas Rule, *European Patent* No. 0,129,990, Sept. 9, 1987.

[22] Angel Gil, *European Patent* No. 0,484,266, May 6, 1992.

[23] Matthew A. Kuchan, Marc L. Masor, Debra L. Ponder, Robin J. Halter, John D. Benson and Gary E. Katz. *United States Patent* No. 6,136,858, Oct. 24, 2000.

[24] Paul Thomas Quinlan, Ian Christopher Chandler. *European Patent* No. 0,496,456, Jul. 29, 1992.

[25] E. Lopez-Huertas, *Pharmacol. Res. 61*, 200 (2010).

[26] J. Romeo, J. Warnberg, E. García-Mármol, M. Rodríguez-Rodríguez, L. E. Diaz, S. Gomez-Martínez, B. Cueto, E. López-Huertas, M. Cepero, J. J. Boza, J. Fonollá, and A. Marcos. *Nutr. Metab. & Cardiovasc. Dis. 21*, 113 (2011).

[27] T. Bourtoom. *Int. Food Res. J. 15*, 237 (2008).

[28] J. W. Rhim, Y. Wu, C.L. Weller and M. Schnepf, *J. Food Sci. 64*, 149 (1999)

[29] M. J. Fabra, P. Talens and A. Chiralt, *J. Food Eng. 85*, 393 (2008).

[30] V. Morillon, F. Debeaufort, G. Blond, M. Capelle and A. Voilley, *Crit. Rev. Food Sci. Nutr. 42*, 67, (2002).

In: Oleic Acid: Dietary Sources, Functions and Health Benefits ISBN: 978-1-62618-332-2
Editor: Luciano Paulino Silva

Chapter 10

APPLICATIONS OF OLEIC ACID ON NANOTECHNOLOGY CONCERNING ENZYME IMMOBILIZATION TECHINIQUES

Eduardo F. Barbosa [*] ***and José L. Cardozo Filho***
Institute of Biology, University of Brasilia, Brazil
Embrapa Genetic Resources and Biotechnology, Brasilia, Brazil

ABSTRACT

This chapter describes a review about oleic acid applications in magnetic systems structured at billionth part of matter, related to enzymes immobilization. Basically, it is shown the oleic acid stabilizing effects over magnetic nanoparticles, mainly those derived from iron. Some oleic acid properties are discussed specifically concerning the reducing effects in magnetic nanoparticles magnetization saturation, as well as size control improvements. Protocols for nanostructures production were gathered to provide compact information about different routes to obtain controllably nanostructured systems, allowing the reader to compare some synthesis strategies. A compilation of studies related to enzyme immobilization (mainly lipases from different organisms) using oleic acid based materials was presented, highlighting the benefits obtained from this technique, such as a better tolerance against adverse catalysis conditions and kinetic parameters. Other oleic acid based materials applications were described, showing the high versatility of these materials. Finally, this chapter points out the use of nanostructured oleic acid based systems as support for enzyme immobilization like an opened research field, which requires extensive investigations to whole clarify its application and potentialities on biotechnology.

INTRODUCTION

Enzyme immobilization techniques have been widely applied and investigated since its discovery, in 1916, by Nelson and Griffin [1]. These techniques present the benefit of employing particular enzyme properties, such as specificity and efficiency, mild reaction

[*] E-mail address: barbosaeduardofernandes@gmail.com

conditions, and also some activity improvements that can be resultant of the micro-environment originated from immobilization process [2]. Systematically, enzyme immobilization can be classified in three groups, according to the linkage or imprisonment established between enzyme and support or carrier: binding to a support (carrier), entrapment (encapsulation), and cross-linking [3]. Moreover, the immobilization method influence and enzyme structure, the supported enzyme preparation properties are guided by the carrier physical-chemical nature.

Nanoparticles have been largely used for enzyme immobilization technology [4]. These materials present specific and unique features, related to the matter organization and arrangement on nanoscale. Some of those characteristics enable the systems to reach enhanced enzymatic performance [4]. Magnetic nanoparticles have been widely applied to enzyme immobilization, possible by offering the advantage of being removed or separated from reaction system just by an external magnetic field application [5]. In this context, this property is very remarkable, allowing a better reaction course control and facilitating enzyme recovery for new cycles of use.

Magnetic nanoparticles in a solvent suspended form can be considered as an unstable system, tending to aggregate or agglomerate, consequently requiring some stabilizing agent [6].

In this context, oleic acid has been described as the most used stabilizer agent. Then, this chapter discusses the oleic acid influence on the surface or inner (including in some preparation step) of nanosized magnetic materials, as well as the stabilization generated by this molecule on the support or carrier components. Furthermore, it also highlights their effects in enzyme immobilization processes and some possible influences related to the catalytic power.

Oleic Acid as a Stabilizing Agent on Nanoparticles Systems

One of the first studies using oleic acid on particles structured in the billionth part of the matter was published in early 80's by Berkowitz et al. [7]. In this seminal work they produced magnetic fluid of iron composites by prolonged grinding in kerosene method followed by the particles covering procedure with oleic acid. These results demostrated a pronounced stabilizing effect caused by oleic acid on magnetic fluid particles [7]. Coating particles with oleic acid increases local anisotropy in magnetic particles to higher values and, consequently, contributes to reduce magnetization saturation events. Since then a massive number of related works emerged on the scientific scenario concerning nanosized magnetic particles or metallic systems nanometrically organized. In the majority of these studies oleic acid stabilizing properties on nanosized systems were highlighted and emphasized [7].

Most of strategies found in literature employing oleic acid as stabilizing agent involve nanosized magnetic particles, particularly those particles constituted by iron. For nanoparticles development and its applications, it is strongly desired particles sizing around 20 nm or less [6]. However, at this scale of matter many problems can arise considering working with magnetic particles. These problems are deeply related to increasing surface/volume ratio energies, becoming a great challenge to produce magnetic particles with desired size and acceptable size distribution without particles aggregation [6, 8]. Some problems described previously resulted from the fact that magnetic nanoparticles have special characteristics in physical and chemical scope.

According to Sohn these characteristics are not proper neither to the atom nor the bulk counterparts [9]. Then, in the same way, any reduction in magnetic particles size to the nanometer scale results in superparamagnetic effect and magnetization quantum tunneling. These phenomena occur possibly due to the fact that each particle could be considered as unique magnetic entity. Consequently nanoparticles tend to aggregate and to reduce its surface energies [8]. Studies conducted by Sun et al. demonstrated that this event occurs partially by virtue of the extensive versatility that MFe_2O_4 composites (ferrites) organized at nanometer-scale can offer to the development and advance of many technologic fields [10]. For example, this composite has been largely used in new technologies development in several areas including electronic, information storage, medical diagnosis and delivery drug based on nanoparticles [10].

Magnetic iron oxide nanosized particles have high reactivity index and in contact with the air these particles can be easily oxidized. One of the most undesirable oxidation effect is correlated to loss of particle magnetism. In this sense some alternatives arise to overcome it. So, increasing particles surface stability by coating them with some compounds (from small molecules to macromolecules) is one way that have been commonly used. A great number of hydrophobic molecules, mainly oleic acid, have been chosen to modify magnetic particles surfaces in synthesis steps. The oleic acid fundamental role in particles stabilizing is related to structural feature, owning a double bond located at position 9 and assuming *cis* conformation. Unsaturation causes a tensioning degree in oleic acid chain and this is responsible for the magnetic particles stabilization [8, 11]. Moreover, critically advantage obtained by using biomolecules like oleic acid arises from the fact that these compounds present good acceptability by human organisms [8]. The main surface modification effect on magnetic particle by oleic acid covering is the layer (or layers) formation around the particles, which reduces its aggregation and optimizes the particles size. For a stable dispersion, it is necessary that repulsive forces are stronger than the attractive ones making the ferrofluid stable. This can be achieved by electrostatic repulsion of similarly charged surfaces that results from an electric double layer creation around the particles, which is dependent of particles dispersion into a polar media, pH, concentration, and ionic strength. At this point it is important to highlight that nanosized magnetic particles application depends on the homogeneity and the particles shape. Moreover, particles chemical and physical stability in diverse systems is very desirable [12, 4].

Magnetic iron oxide magnetite (Fe_3O_4) is the strongest paramagnetic compound encountered in nature [13]. Magnetite (Fe_3O_4) has been widely employed to develop different type of products and consequently many efforts are made to enlarge the knowledgement about the particles production based on this compound. In this sense there are a large number of methods to produce these nanoparticles, for instance: co-precipitation, micro-emulsion, electrochemical synthesis, pyrolysis, and hydrothermal synthesis [14]. Many examples about oleic acid applicability in nanotechnology developments can be find in literature. Although most of published material using oleic acid as stabilizing agent refers to iron nanoparticles, there are an increasing number of studies concerning several nanoparticles synthesis constituted by different inorganic compounds. Puntes et al. applied oleic acid with trioctylphosphine oxide to control cobalt nanorods growth rates [12]. Materials based on

CdSe/CdS have been studied and developed with use of oleic acid to synthesize the CdSe /CdS quantum dots [15, 16]. Studies involving TiO_2 doping have been improving properties of this material and others that results in enlargement its photoactivity capacity [17]. Semiconductor materials represent a potential field concerning oleic acid use, when aiming to obtain controlled size and shape distributions in nanocrystals for some materials, as for cadmium, for instance [18].

Lattuada and Hatton developed a study functionalizing different nanoparticles groups using oleic acid and other compounds to stabilize particles and control their size homogeneity [19].

In parallel to all of these benefits showed previously, systems containing oleic acid has been also applied as stabilizing agent to several other materials, including in enzyme immobilization processes and this will be discussed in details on the next topic.

Applications in Biotechnology and Enzyme Immobilization Using Oleic Acid Based Materials

There is an extensive number of methods to build nanosized magnetic particles coated with oleic acid currently available, which allows glimpsing the use of these particles as support to immobilize enzymes. The application of these structures on enzyme immobilization studies can be still considered as a field in development, since there is a no large amount of studies and information related to this subject.

Following, we present different strategies to obtain magnetic nanoparticles containing oleic acid, which own specific properties and applications. These methods allow controlling the particles size by diverse condition combinations. Sun et al. as an example, achieved particles with high degree of size homogeneity using Fe_3O_4 [6]. The particles size was observed ranging from 4 to 16 nm, considering the oleic acid presence on these particles surface. These results were obtained using the two subsequent protocols:

Protocol 1:

Iron (III) acetylacetonate ($Fe(acac)_3$) (2 mmol) was mixed in phenyl ether (20 mL) with 1,2-hexadecanediol (10 mmol), oleic acid (6 mmol), and oleylamine (6 mmol) under nitrogen and was heated (265°C) to reflux for 30 min. After cooled at room temperature, the dark-brown mixture was treated with ethanol under ambient air, and a dark-brown material was precipitated from the solution. The product was dissolved in hexane in the presence of oleic acid and oleylamine and reprecipitated with ethanol to give 4 nm Fe_3O_4 nanoparticles.

Protocol 2:

Mixing and heating (265°C) 62 mg of 8 nm Fe_3O_4 nanoparticles with 2 mmol of $Fe(acac)_3$, 10 mmol stearyl alcohol, 2 mmol of oleic acid, and oleylamine (2 mmol) led to 12 nm Fe_3O_4 nanoparticles, while changing the mass of seeds into 15 mg led to 16 nm Fe_3O_4 nanoparticles.

Despite the theoretical stability attributed to oleic acid, on that study it still was observed some Fe_3O_4 aggregated particles [6].

About iron nanoparticles synthesis methods involving thermal decomposition using oleic acid, Hyeon et al. used iron pentacarbonyl oxidation and achieved maghemite particles (γ-Fe_2O_3) with size varying from 4 to 16 nm and dimension uniformity among the particles [20]. The control over particles size was made by molar ratio variation between $Fe(CO)_5$ and oleic acid. When the molar ratio $Fe(CO)_5$ /oleic acid was 1:3 the obtained particles size was around 11 nanometers. Ratios of 1:1 and 1:2 lead to particles with sizes from 4 to 7 nm. To obtain particles with sizes between 11 and 16 nm the molar ratio $Fe(CO)_5$ /oleic acid was increased to 1:4 and the protocol used is detailed below [20]:

Protocol 3:

A volume of 0.2 mL of $Fe(CO)_5$ (1.52 mmol) was added to a mixture containing 10 mL of octyl-ether and 1.28 g of oleic acid (4.56 mmol) at 100°C. The mixture was heated to reflux and kept at that temperature for 1 h. The resulting black solution was cooled to room temperature and 0.34 g of dehydrated trimethylamine oxide ($(CH_3)_3NO$) (4.56 mmol) was added. The mixture was heated to 130°C under an argon atmosphere and maintained at this temperature for 2 h, whereupon it formed a brown solution. The reaction temperature was slowly increased to reflux and maintained for further 1 h. The solution was then cooled to room temperature, and ethanol was added to yield a black precipitate, which was separated by centrifugation.

Yu et al. used methods to build dumbbell-like Au-Fe_3O_4 nanoparticles with adjustable sizes in the range 3 to 14 nm [14]. Basically nanoparticles were built taking in consideration two distinct routes: chloroauric acid addition ($HAuCl_4$) in solution containing iron pentacarbonyl $Fe(CO)_5$ (Protocol 4), or starting from pre-made Au nanoparticles (Protocol 5). These methods are described below:

Protocol 4:

Oleic acid (6 mmol), oleylamine (6 mmol), 1,2 hexadecandiol (10 mmol) and 20 mL 1-octadecene (ODE) were mixed and stirred under a gentle flow of nitrogen at 120ºC for 20 min. Then under a blanket of nitrogen, 0.30 mL $Fe(CO)_5$ (2 mmol) was injected into the solution. After 3 min, the deaerated gold precursor solution consisting of 40 mg $HAuCl_4{\bullet}(H_2O)_3$ (0.1 mmol), 0.5 mL oleylamine (1.5 mmol) and 5 mL ODE was injected into the solution. The solution turned to dark red instantly after this procedure, indicating the formation of gold nanoparticles. The mixture was heated to reflux (~310°C) for 45 min, cooled down to room temperature, and exposed to air for an extra hour to ensure the formation of Fe_3O_4. The particles were precipitated out with isopropanol (~40 mL) addition followed by centrifugation. The precipitate was re-dispersed into hexane in the presence of ~0.05 mL oleylamine and centrifuged again to remove any undispersed materials. The dumbbell nanoparticles were precipitated out by adding ethanol and re-dispersed in hexane in the presence of ~0.05 mL oleylamine, giving a dark red brown dispersion. A little extra of oleylamine was necessary to ensure long term stability of the dispersion. Such dispersion could be stored under ambient condition for over one month without any noticeable agglomeration. The yield of the synthesis was dependent on the final purification step and was about 50-70%.

Protocol 5:

Au nanoparticles with average size around 8 nm were made as follow: 1.0 g $HAuCl_4(H_2O)_3$ (2.5 mmol) was added to 100 mL tetralin, followed by 10 mL oleylamine (30 mmol) to form a red solution. The solution was heated at 65°C for 5 h and cooled to room temperature. Ethanol was added to the solution, and gold particles were separated by centrifugation, washed by ethanol, and redispersed in hexane. To produce Au-Fe_3O_4 a solution of 1mL oleic acid (3 mmol) in 20 mL octadecene was heated at 120°C for 20 min under a N_2 flow. Then under a blanket of N_2, 0.15 mL $Fe(CO)_5$ was injected to the solution. After 5 min of stirring, 0.5 mL oleylamine was injected to the reaction mixture, followed by addition of 2 mL of 8 nm Au colloidal dispersion previously made (ca. 20 mg Au). The solution was heated to reflux (ca. 310°C) for 45 min. After cooled down to room temperature, the particles were separated by adding isopropanol, centrifuged and redispersed in hexane.

An important observation made by Yu et al. is that Au-Fe_3O_4 particles with size up to 8 nm and Fe_3O_4 particles up to 20 nm were readily stabilized in hexane [14]. Wang *et al* presented a Fe_3O_4/carbon nanocomposition formulation using Fe_3O_4 nanoparticles coated with oleic acid and heating a glucose aqueous solution as description that follows below [21]:

Protocol 6:

Oleic acid-stabilized Fe_3O_4 magnetic fluid was prepared by firstly adding 3 mL of oleic acid into 220 mL of the aqueous solution containing 10 g/L Fe_3O_4 nanoparticle at 80°C under vigorous stirring, and then 1 mL of $NH_3OH\_H_2O$ (25%) was added dropwise to the above solution. After that, the reaction mixture was stirred for another 0.5 h at 80°C. Finally, oleic acid-stabilized Fe_3O_4 nanoparticle was separated from the mixture by ultracentrifugation and washed with water three times and with ethanol two times. To prepare Fe_3O_4/carbon nanocomposition the reactant mixtures were prepared by dissolving glucose in the above aqueous solution containing oleic acid-stabilized Fe_3O_4 nanoparticles. In the typical reaction, 4 g of glucose was dissolved in 40 mL of 2.5 g/L oleic acid stabilized Fe_3O_4 nanoparticles aqueous solution. The reactant mixture was placed in a 50 mL Teflon-sealed autoclave and heated at 170°C for 3 h. Finally, the product was isolated from the mixture by the external magnetic field and it was cleaned by three cycles of separation/washing/redispersion in water and in the mixture of ethanol/chloroform (1:1).

Vidal-Vidal et al. reported the maghemite nanoparticles synthesis coated using a microemulsion method of cyclohexane/polyoxyethylene 10 oleoyl ether/aqueos phase containing oleylamine as precipitant agent [22].

Protocol 7:

The aqueous phase was previously prepared by a mixture of 1 M in Fe(III), 0.5 M in Fe(II) and 0.1 M in HCl to avoid Fe(II) oxidation. For synthesis, 250 mL of a microemulsion containing a volume ratio 90/7/3 of cyclohexane/Brij-97/aqueous phase was prepared. The mixture of components was stirred and purged with N_2 at room temperature during 10 min. Then, samples were placed in a thermostatic bath at 50°C with magnetic stirring. The injection of oleylamine was carried out waiting 5 min after the microemulsion was formed. Oleic acid coated particles were synthesized injecting 13.8 mL of oleylamine diluted in 10 mL of cyclohexane, at 50°C. After waiting 15 min, 150 mL of the synthesis product was separated. A mixture of 2.40 g of oleic acid and 3.24 g of oleylamine diluted in cyclohexane was added to this part and stirred for 30 min at 50°C. The particles obtained in both parts of this synthesis were then flocculated with ethanol, separated and redispersed two times in heptane. Finally, particles were dried under vacuum at 30°C. In this specificcase, oleylamine could be replaced by oleic acid.

Before applying oleic acid as a stabilizing agent on enzyme immobilization, some studies concerning lipase immobilization demonstrated some benefits of using hydrophobic carriers to enhance its activity. The first studies using oleic acid in enzymatic systems did not consider the nanostructured aspects of the matter, but it evaluated as a natural system works (naturally nano-organized) and applied some features of it according to their interests. Aires Barros et al. study represents one of the most important studies related to the evaluation of the oleic acid influence and other fatty acids on *Saccharomyces bayanos* enzymes activities, evaluating the ethanol production by an extractive fermentation system containing immobilized cells at high sugar concentrations [23]. Al Duri and Young conducted experiments using lipases immobilization concerning hydrophilic and hydrophobic surfaces as supports [24]. They concluded that support physical and structural properties and lipase physical and chemical features make possible the equilibrium behavior of lipase/support materials in immobilization processes. Other important conclusion was the considerable adsorption capacity of both support materials. Nevertheless, each material presented different adsorption models, determining the activity efficiency of immobilized lipases. For instance, the hydrophilic support presented higher lipase adsorption capacity, but it did not improve the lipase activity and efficiency. This probably happen because the immobilization produced equilibrium isotherms consisted of two plateaus, being explained by the bi-modal adsorption model proposed by Fair and Jamieson [25]. It is important to be aware that regular and monolayer adsorption is wished, not heterogeneous and multilayer ones, or ever cluster formation [24]. Oliveira et al. also pointed out some benefits of immobilizing lipase onto hydrophobic carriers [26]. They produced a perforated rotating disc bioreactor, developed to perform the ethanol esterification with oleic acid, catalyzed by a lipase from *Rhizomucor miehei* immobilized by adsorption onto a hydrophobic support: Accurel EP700.

Guo et al. reported a preparation and characterization of immobilized *Candida cylindracea* lipase on magnetic hydrophobic microspheres (8-34 μm), using co-polymerized vinyl acetate and divinyl benzene with oleic acid-stabilized magnetic colloid core [27]. This carrier presented remarkable enhanced enzyme loading power (6.750 IU/g of carrier), comparing to other previous studies. Beyond its superparamagnetic characteristics, which facilitate sedimentation and reaction course control, immobilization process increased thermal stability and reusability, suggesting an interfacial activation after immobilization [27]. Other studies demonstrated that *Candida cylindracea* lipase can be activated by interactions with a hydrophobic interface by changing conformation in its secondary structure [28]. When hydrophobic interface is absence, some enzyme portion covers its active site, making it unreachable to substrates [29]. The enzyme interaction with hydrophobic interface can induce conformational changes activating the lipase [30].

The hydrophobic microenvironment establishment can offer a preferable way to activate the lipase. Thus, lipase immobilization is projected to supply this hydrophobic microenvironment [31]. Lei *et al* synthesized magnetic microspheres by the suspension polymerization of glycidylmethacrylate, methacrylicacid, and divinylbenzene in the presence of oleic acid-coated Fe_3O_4 nanoparticles [32]. Triacylglycerol lipase from porcine pancreas was covalently immobilized on the magnetic microspheres via other entrapment methodology the active epoxy groups. It presented better enzyme loading (39 ± 0.5 mg/g support), and resulted in higher optimum temperature compared with those of free lipase, exhibiting superior thermal and pH stability and excellent reusability.

Mahmood *et al* reported a study correlated to immobilization by adsorption of lipase from *Candida cylindraces* via strong hydrophobic interactions [33]. They used an oleic acid-pluronic L-64 block copolymer coated iron oxide nanoparticles as supports for enzyme immobilization. The activity of the immobilized lipase maintained at levels as high as 90% of its original activity for more than seven reuses. Stabilizing features are commonly showed by enzyme immobilization. There is a vast literature about immobilization technology increasing activity tolerance against adverse conditions, such as pH, ionic strength, temperature, and inappropriate solvents. In some cases, kinetic constants can be also improved, together with storage stability and reusability [2]. Another work that emphasized the oleic acid importance on enzyme systems was presented by Hama et al. [34]. They tested *Rhizopus oryzae* lipase activity of immobilized cells, with no pre-treatment, but containing oleic acid enriched membranes for biodiesel fuel production. Oleic acid enriched membranes showed higher initial methanolysis activity, possibly due to the increased permeability and rigidity [34].

An interesting system containing micron-sized magnetic poly (methyl methacrylate-divinylbenzene-glycidyl methacrylate) microspheres was investigated by Ma Z. et al. [35]. This system was prepared by a modified suspension polymerization in the presence of oleic acid-coated magnetite nanoparticles. They functionalized the magnetic microspheres by reacting epoxy groups with ammonia solution to provide amino groups. After activation with glutaraldehyde, bovine serum albumin (BSA) was covalently immobilized on these magnetic microspheres, aiming to use BSA as a model protein to covalent immobilize on the magnetic microsphere. They highlight that this technique could be applied to enzymatic systems, due to the positive features of using covalent bond to immobilize selectively proteins, over optimum ionic strength, pH, and protein concentration conditions to obtain higher yields [34].

Yong *et al* described a study concerning *Candida rugosa* lipase covalently immobilized on the hydrophilic magnetic microspheres via the active epoxy groups [36]. They prepared magnetic microspheres by suspension polymerization of glycidyl methacrylate, ethylene glycol dimethacrylate and vinyl acetate in the presence of oleic acid-coated magnetite, following by alcohol treatment to increase their hydrophilicity. Beyond enhanced resistance against pH and temperature inactivation in comparison to free lipase, this system presented better thermostability and reusability. The Michaelis-Menten constant value (K_m) for immobilized (1.03 mg/mL) was higher than free lipase (0.52 mg/mL), while the value of the maximum rate achieve by the enzymatic system (V_{max}) was smaller for the immobilized lipase [36].

In addition to enzyme immobilization, other interesting applications of oleic acid based nanosystems can be pointed. For instance, Andhariya *et al* described a study of a new drug delivery for doxorubicin using biodegradable thermoresponsive polymeric magnetic nanoparticles [37].

Nagarajan Sounderya *et al* presented a study of imaging gap junctions with silica-coated up conversion nanoparticles [38]. Ma Z. *et al* used micro-sized magnetic poly vinyl acetate-divinylbenzene prepared in presence of oleic acid coated nanoparticles, modified by ester exchange reaction, as a magnetic chelator for high capacity immobilized metal affinity adsorption of protein by cerium initiated graft polymerization [39]. Whelehan *et al* used a porous hydrogel membrane to form liquid-core microcapsules which can be used to extract some pharmaceutical compounds for water decontamination [40].

CONCLUSION

This chapter evaluated structures based on oleic acid components and its features. It was showed how the oleic acid molecule acts as a stabilizing agent, mainly in ferric magnetic nanoparticles. A set of protocols to obtain nanostructures was reported, aiming to show some routes that could control these nanosystems, and to compare the synthesis strategies referring to size range and other properties. A collection of studies concerning enzyme immobilization (mainly lipases from different organisms) and oleic acid based materials was discussed. In an overview, stabilizing features were showed by immobilized enzyme, increasing its tolerance against adverse catalysis conditions, such as pH, ionic strength, temperature, and inappropriate solvents. In this context, kinetic constants can also be improved, together to storage stability and reusability. Finally, other oleic acid based materials applications were presented, showing the high versatility of these materials. This chapter, represents a short review of the *state-of-art* about some advances reached by oleic acid-based materials and its applications, revealing this scenario as an opened research field.

ACKNOWLEDGMENTS

The authors acknowledge Dr. Thales Lima Rocha by the English revision on this chapter.

REFERENCES

[1] J. M. Nelson, E. G. Griffin, *J. Am. Chem. Soc. 38,* 1109 (1916).
[2] R. A. Sheldon, *J. Chem.*57, 281 (2004).
[3] R. A. Sheldon, *Adv. Synth. Catal.* 349, 1289 (2007).
[4] J. Kim, J. W. Grate, P. Wang, *Chem. Eng. Sci.* 61, 1017, (2006).
[5] A., Dyal, K. Loos, M. Noto, S.W. Chang, C. Spagnoli, K. Shafi, A. Ulman, M. Cowman, R. A. Gross, *J. Am. Chem. Soc.* 125, 1684 (2003).
[6] S. Sun, H. Zeng. *J. Am. Chem. Soc.,* 124, 8204 (2002).
[7] A.E. Berkowitz, J.A. Lahut, C.E. VanBuren, *IEEE Transactions on magnetics*, 16, 184 (1980).
[8] W. Wu, Q. He, C. Jiang, *Nanoscale Res. Lett.* 3, 397 (2008).
[9] B.H. Sohn, R.E. Cohen, G.C. Papaefthyrniou, *J. Magn. Magn. Mat.*, 182, 216 (1998).
[10] S. Sun, H. Zeng, D. B. Robinson, S. Raoux, P. M. Rice, S. X. Wang,G. Li, *J. Am. Chem. Soc.,*126, 273 (2004).
[11] K. Gupta, M. Gupta, *Biomaterials*, 26, 3995 (2005).
[12] V. F. Puntes, K. M. Krishnan, A. P. Alivisatos, *Science*, 291, 2115 (2011).
[13] R. J. Harrison, R. E. Dunin-Borkowski, A. Putnis, *PNAS*, 99, 16556 (2002).
[14] H. Yu, M. Chen, P.hilip M. Rice, S. X. Wang,R. L. White, S. Sun, *Nano Lett.* 5, 379 (2005).
[15] J. J. Li, Y. A. Wang, W. Guo, J. C. Keay, . *J. Am. Chem. Soc.,* 125, 12567 (2003).

[16] H. Y. Kang, 'A Review of the Emerging Nanotechnology Industry: Materials, Fabrications, and Applications'. <http://www.dtsc.ca. gov/Technology Development/ Nanotechnology/upload/Review_of_Emerging_Nanotech_Industry.pdf>
[17] N. Banerjee, *Nanotechnology, Science and Applications*, 4, 35 (2011).
[18] Y. Mao, Tae-Jin Park, F. Zhang, H. Zhou, S. S. Wong, *Small*, 3, 1122 (2007).
[19] M. Lattuada, T. A. Hatton, *Langmuir*, 23, 2158 (2007).
[20] T. Hyeon, S. S. Lee, J. Park, Y. Chung, H. B. Na, *J. Am. Chem. Soc.* 123, 12798 (2001).
[21] Z. Wang, H. Guo, Y. Yu, N. He, *J. Magn. Magn. Mat.*, 302, 397 (2006).
[22] J. Vidal-Vidal, J. Rivasb, M.A. Lopez-Quintela, *Colloids and Surfaces A: Physicochem. Eng. Aspects,* 288, 44 (2006).
[23] M. R. A. Barros, J. M. S. Cabral, J. M. Novais, *Biotech. Bioeng.* 29, 1097 (1987).
[24] B. Al-Duri, Y. P. Yong, *Bioch. Eng. J.*, 4, 207 (2000).
[25] B.D. Fair, A.M. Jamieson, *J. Colloid and Interface Sci.* 77, 525 (1980).
[26] A.C. Oliveiraa, M.F. Rosa, M.R. Aires–Barros, J.M.S. Cabral, *Enz. Mic. Tech.*, 26, 446 (2000).
[27] Z. Guo, S. Bai, Y. Sun, *Enz. Mic. Tech.,* 32, 776 (2003).
[28] L. Brady, A. M. Brzozowski, Z. S. Derewenda, E. Dodson, G. Dodson, S. Tolley, et al. *Nature*, 343, 767 (1990).
[29] P. Grochulski, Y. Li, J. D. Schragm, F. Boutthilier, P. Smith, D. Harrinson, et al. *J. Biol. Chem.* 268,12843 (1993).
[30] P. Grochulski, J. Li, J. D. Schrag, M. Cygler, *Protein Sci.*, 3, 82 (1994).
[31] R. Fernandez-Lafuente, P. Armisen, P. Sabuquillo, G. Fernandez-Lorente, J. M. Guisan, *Chem. Phy. Lipids,* 93, 185 (1998).
[32] L. Lei, Y. Bai, Y. Li, L. Yi, Y. Yang, C. Xia, *J. Magn. Magn. Mat.*, 321, 252 (2009).
[33] Mahmood, C. Guo, H. Xia, J. Ma, Y. Jiang, H. Liu, *Ind. Eng. Chem. Res.* 47, 6379 (2008).
[34] S. Hama, H. Yamaji, M. Kaieda, M. Oda, A. Kondo, H. Fukuda, *Bioch. Eng. J.*, 21, 155 (2004).
[35] Z. Y. Ma, Y. P. Guan, X. Q. Liu, H. Z. Liu, *Langmuir*, 21, 6987 (2005).
[36] Y.Yong, Yong-Xiao Bai, Yan-Feng Li, L. Lin, Yan-Jun Cui, Chun-Gu Xia, *Process Biochemistry*, 43, 1179 (2008).
[37] N. Andhariya, B. Chudasama, R. V. Mehta, R. V. Upadhyay, *J. Nanopart. Res.*, 13, 1677 (2011).
[38] S. Nagarajan, Z. Li, V. Marchi-Artzner, F. Grasset, Y. Zhang, *Med. Biol. Eng. Comput.* 48, 1033 (2010).
[39] Z. Ma, Y. Guan, X. Liu, H. Liu, *Polym. Adv. Technol.*, 16, 554 (2005).
[40] M. Whelehan, U. von Stockar, I. W. Marison, *Water Research*, 44, 2314 (2010).

In: Oleic Acid: Dietary Sources, Functions and Health Benefits ISBN: 978-1-62618-332-2
Editor: Luciano Paulino Silva

Chapter 11

OLEIC ACID AND CANCER: FROM THERAPEUTIC TO CO-ADJUVANT AGENTS IN ANTICANCER APPROACHES

Graziella A. Joanitti*

Institute of Biological Sciences, University of Brasilia, Brazil
Embrapa Genetic Resources and Biotechnology, Brasilia, Brazil

ABSTRACT

The heterogeneity and versatility nature of cancer emphasize why cancer treatment is considered a challenging task. The search for new tools has not only been based on the discovery of new molecules, but also on the discovery of new functions for known molecules. Oleic acid is one of those examples. Oleic acid is a monounsaturated (C18:1, n-9) fatty acid found in animal and plant fats often included in the human diet and is already known for its nutritional and structural properties. However, this fatty acid has also been extensively studied for its anticancer properties. The aim of this chapter is to gather the current knowledge regarding the effects of oleic acid in cancer and to provide a review of its versatile applications as an anticancer molecule itself, a co-factor/stabilizing agent, a skin penetration enhancer molecule, and as nanoparticle components aiming to improve, complement, and enable developing anticancer therapies.

INTRODUCTION

Cancer is a multistep process basically characterized by genetic mutations and epigenetic changes resulting in increased cell proliferation/growth rates, invasion, and ultimately spreading of these cells to other tissues in the body. The complexity nature of cancer challenges the establishment of appropriate therapeutic treatments and contributes to maintain the status of this pathology as the leading cause of death worldwide.

* E-mail address: bygra1@gmail.com

Several efforts have been made in order to improve the knowledge about the biology of cancer and to identify factors able to negatively affect this pathology. Dietary habits have been suggested to be one of the factors influencing cancer occurrence/progress. This evidence have triggered close investigations of food compounds suggesting that molecules such as some secondary compounds, proteins, vitamins, and fatty acids could act as anticancer molecules in different stages of the disease.

Oleic acid is a monounsaturated (C18:1 n-9) fatty acid found in animal (e.g. chicken) and plant (e.g olives) fats often included in the human diet; with consumptions rates varying according to the world region. Its anticancer properties have been investigated and widely discussed from a preventive to a co-adjuvant perspective. In this context, the aim of this chapter is to gather the current knowledge regarding the effects of oleic acid in cancer and to provide a review of its role as an anticancer molecule itself and as a cofactor/stabilizing agent to improve, complement, and enable other developing anticancer therapies.

A Brief Review on the Complex Nature of Cancer

Cancer is considered the leading cause of death in economically developed countries and the second leading cause of death in developing countries. Lung, breast, colorectum, stomach and prostate cancer, listed in descending order, are among the most prevalent cancer types worldwide; accounting for approximately 50% of all diagnosed cases [1]. In 2008, 7.6 million cancer deaths were estimated to have occurred (around 13% of all deaths) and these rates are projected to continue rising, with an estimated 12 million deaths in 2030 [1, 2, 3].

Genetic mutations acquired and accumulated by cells from any human tissue is potentially sufficient to start a cancer development. DNA mutations occurring due to inherent genetic features comprise less than 15% of all cancer cases.

The majority of cancer incidence is attributed to several environmental factors and lifestyle habits/behavior which are able to induce not only genetic, but also epigenetic (*e.g.* methylation) changes in the DNA. Those mutations can affect the expression of oncogenes and tumor suppressor genes leading to an impairment of the intracellular circuits responsible to regulate and control cell homeostasis such as proliferation [4].

Some would even classify a tumor as a new tissue since it comprises cells with heterogenic characteristics. Cancer features can be summarized in established and emerging hallmarks corresponding to capabilities acquired by the cells during the multistep process of human tumor pathogenesis [4]: sustaining proliferative signaling; evading growth suppressors; invasion and metastasis; replicative immortality; inducing angiogenesis; resisting cell death; deregulating cellular energetic; and avoiding immune destruction. Each of these categories represents a potential target for anticancer therapies; and, indeed, several drugs have been tested against them. However, some approaches have not resulted in the expected outcome. For example, it was reported that the use of potent angiogenesis inhibitors induced an efficient reduction in the tumor neovascularization. Nevertheless, the tumor counteracted to this therapy by triggering other pathway - invasiveness and metastasis [4, 5]. Thus, cancer cells are able to shift or change signaling pathways to compensate the impairment of one of the hallmarks induced by the treatment [4].

These results reflect the heterogeneity and versatility nature of cancer and also emphasize why cancer treatment is considered a challenging task. Therefore, there is a need to discover/improve tools and approaches available for cancer prevention and treatment.

The search for new tools has not only been based on the discovery of new molecules, but it has been based on the discovery of new functions for known molecules as well. Even classic classes of macromolecules (*e.g.* nucleic acids, lipids) have been revisited and tested for their anticancer properties. Oleic acid is one of those examples. It is a monounsaturated fatty acid, already known for its nutritional and structural properties that have also been studied for anticancer therapies showing encouraging results which will be described below.

Oleic Acid as a Preventive Agent and a Predictive Biomarker

During the last few decades, advances in combining screening and treatment approaches have effectively reduced the mortality of some cancers; however, limitations in treatment, accessibility and clinical interventions for some other cancers still remain to be addressed. Considering the complexity of cancer biology and the difficulties to establish an appropriate therapy, several researches have suggested that prevention approaches might be reasonable options for reducing the increasing cancer rates [6].

It is known that cancer incidence varies dramatically among populations worldwide. This pattern could be attributed to specific genetic factors present in each population. Nevertheless, epidemiologic studies have reported markedly changes in cancer rates among populations migrating from low- to high-incidence countries [7]. Within one to three generations, cancer rates of the migrant population can acquire cancer rates similar to the rates of their new homeland [8-10].These data indicates that other factors, rather than genetic, are able to influence cancer rates. It means that a substantial proportion of cancer cases could be prevented through the identification of approaches to limit the exposition, change the frequency or offer a protection against these factors. It is known that some aspects of the environment and lifestyle in low-incidence countries are dramatically distinct from those countries with high cancer rates, especially regarding diet [11].

The influence of diet in cancer rates has been widely investigated. Interestingly, it has been suggested that 20% to 60% of cancers could be preventable by altering the diet [8]. Among several dietary components that could modulate cancer risk, oils and fats has been extensively investigated providing evidence that exogenously dietary fatty acids may play an important role in cancer evolution and/or progression [12]. Mediterranean countries, which are known for their low-cancer incidence, have a distinct diet when compared to high-incidence rate countries. Specific food consumption patterns typical of Mediterranean regions (Greece, Spain, and Southern Italy) in the early 1960s is known as the "Traditional Mediterranean Diet" [13]. This diet is characterized by a high intake of legumes, fruits, vegetables, nuts, fish, small quantities of red meat, and moderate alcohol consumption. The intake of fat is high; with the proportion of monounsaturated fatty acids corresponding to approximately twice the saturated fatty acids [13, 14].

Olive oil is the main source of fat in the Mediterranean Diet and studies based on the amount of its consumption have supported its chemoprotective properties in cancer [15]. It has been described that the mean yearly intake of olive oil in Mediterranean countries is approximately 15 times higher than in other countries [16] and that countries with high consumption of olive oil have lower mortality rates from breast and colorectal cancer than those with low olive-oil consumption [17, 18]. Several epidemiologic studies have suggested that a high intake of olive oil is related to a lower risk of cancer incidence, in particular breast, stomach, ovary, colon, and endometrium cancers [12], emphasizing its role in cancer prevention. In recent studies from Southern Europe, consumption of olive oil was inversely related to breast cancer risk [19, 20].

These data have triggered efforts in the identification of olive oil components and their role in cancer protection. The fat content of olive oil is composed of 6-11.5% polyunsaturated fatty acids (PUFA), 13-16% of saturated fatty acids, and 72-75% of monounsaturated fatty acids (MUFA). The major fatty acid present in olive oil is predominantly oleic acid, a monounsaturated ω-9 fatty acid (OA; cis-C18:1, n-9) [12]. Prevention properties observed for olive oil intake have been attributed to OA, since this fatty acid is its main component, and have stimulated a renewed interest in OA anticancer properties [12, 21].

Although some epidemiologic studies have shown potential protective effects of OA, other studies failed to find a correlation of OA intake (present in olive oil) and cancer prevention. Moreover, it has also been reported that OA might have the opposite effect, in other words, increase cancer risk [22]. For example, case-control studies indicated no correlation of OA intake and breast cancer; while cohort studies showed that OA could be involved in increased breast cancer risk [23].

Altogether, data regarding the effects of OA in cancer have shown inconsistent results. Epidemiological studies are important and provide useful information about the association of diet and disease [23]. Nevertheless, it is important to highlight that several factors may influence in their analysis. First factor is the difference in adjustment for confounding factors known to be related to both cancer risk and fatty acid, such as obesity, weight loss, tobacco and alcohol [23]. Other potential source of variability is the estimation of the amount of fatty acid intake. In these studies, food questionnaires, dietary record, and dietary interview are used to measure amounts of fatty acid consumption indirectly [24]. It is difficult to convert food items into their fatty acid content in particular because their fatty acid content can vary with seasons over the year, cooking methods, and/or industry supplies [11, 23]. The lack of food composition tables specifying fatty acid content also contribute for an underestimation of the fatty acid exposure of each studied subject [11, 23]. Thus, more detailed studies are needed to control for variables that may seriously confound conclusions [11].

In order to overcome these limitations, the use of biomarkers for fatty acids has been suggested as an alternative to access bioavailable and post-absorptive amounts of these compounds more accurately [25]. Fatty acid composition of adipose tissue, erythrocyte membrane, serum, and/or plasma phospholipids have been used in epidemiological studies as such markers of dietary exposure and also of cancer diagnosis/progress [26]. Wood *et al* studied the fatty acid profile of phospholipids in circulating red blood cells membrane in patients with cancer in different stages of progression. The relative increase in the content of OA along with a decrease in stearic acid (C18:0), which is the saturated counterpart of OA, was observed in patients with solid tumors or chronic leukemia. Moreover, higher ratios of OA/stearic acid were observed in patients with advanced cancer suggesting that this biomarker could indicate the degree of malignancy [27]. Similar ratios were observed in lymphoma and malignant liver neoplasms [28].

Nevertheless, other works have reported that the ratio of OA/stearic in the erythrocyte membranes is not related with marker of tumor malignancy [29, 30]. No differences were found when the ratio of OA/stearic was compared with patients with breast cancer in different stages and normal volunteers [29]. A similar pattern was observed in patients with colorectal cancer. No differences in the ratio of OA/stearic was observed in patients with or without cancer even when considering other factors such as age, sex, degree of tumor differentiation, or recurrence [30].

These different conclusions exposed above might be related to the type of cancer since OA/stearic acid ratio was found to be significant in some tumors. However, it should be emphasized that recent dietary intake can alter the pattern of fatty acids in plasma and membrane phospholipids, thus, may not reflect the overall pattern of fatty acids intake over a long period of time. As an alternative, fatty acids composition of adipose tissue seem to be a more interesting marker of long-term dietary intakes since the half-life of fatty acids in this tissue is approximately 2 years [31, 32]. Studies aiming to correlate breast cancer risk through the analysis of fatty acid composition in serum or erythrocytes membrane phospholipids were inconclusive [18]. On the other hand, by the analysis of fatty acids profile in adipose tissues, it was found that an increased on cis-monounsaturates content, including OA, along with a low n-6/n-3 PUFA ratio were associated with protective effects against breast cancer, independently of age and body mass index [11]. These data also show variations in the ratios of several fatty acids and not only in the OA/stearic acid; suggesting that a single fatty acid ratio cannot be considered as an independent biomarker for cancer [11, 12].

Although investigation of the lipidome aids in the conclusion of epidemiological studies and can be an auxiliary index to study cancer progress, it should be stressed that the validation of benefits or harms of an specific fatty acid intake requires a parallel experimental animal study [11]. By this approach, it is possible to control some variables that may interfere in the lipid profile and have a better comprehensive knowledge about of the effects of fatty acids in cancer incidence and progression.

Several animal studies have been conducted in order to evaluate the effects of OA in cancer incidence. Animal studies with OA have been implicated in a decrease in the number of aberrant crypt *foci* at the early stages of tumor development as well as an inhibition of colorectal tumor formation [33, 34]. Comparisons of OA intake (present in olive oil) with other oils presenting higher content of saturated fatty acids showed that rats fed a diet containing 23% olive oil had a lower tumor incidence, fewer tumors per animal, and a longer latent period than in animals fed either the safflower oil or corn oil diet [35]. It has been reported that monounsaturated fatty acids, in particular OA, have been shown to protect against breast cancer, whereas saturated fatty acids were associated with an increase of breast cancer risk in experimental animal studies [36]. Similarly, it was reported that a diet high in oleic acid-rich olive oil reduced the progression of chemically induced rat mammary carcinomas [35]. OA has been described to be also implicated in the prevention of post-operative cancer recurrence. After the complete surgical removal of tumors in mice with spontaneous mammary carcinomas, mice received *i.p.* injections of OA. The results showed that animals treated with OA had lower number of new recurrences of primary tumors and post-operative metastasis. In addition, OA also showed to reduce the effectiveness of tumor grafts re-implantation approaches [37].

Conversely, other animal studies have shown none or an opposite effect of OA in cancer. When chemically induced mammary carcinogenesis rats were fed with a diet containing 20% of olive (high amount of unsaturated fatty acids), sunflower or corn oil (high amounts of saturated fatty acids), no differences in tumor incidence and multiplicity were observed among the experimental groups [38]. This pattern has also been examined in other models, including the development of spontaneous mammary tumors in mice and the growth of transplantable mammary tumors in rats [39, 40]. Furthermore, a meta-analysis of cohort studies, in conjunction with experimental animal studies, have suggested that OA was associated with an increased risk of breast cancer [23, 38, 41].

Lasekan *et al* employed an interesting experimental design to address the effect of both monounsaturated and polyunsaturated fatty acids in breast cancer risk. Chemically induced mammary carcinogenesis rats were fed a 20%-fat diet containing a high-C18:2 n-6 safflower oil, a high- C18:1 safflower oil, or olive oil. Olive oil and the high C18:1 safflower oil had similar concentrations of monounsaturated fatty acids (> 70%), but different C18:2 n-6 contents: 5% and 17% respectively. The results showed that animals fed with olive oil had a longer tumor latency period, a lower tumor incidence, and fewer tumors per rat than in those fed the other two types of fat. When the same amount of C18:2 n-6 were added in the olive oil diet, its protective effect was abolished. Therefore, low amounts of C18:2 n-6 content in olive oil, and not the high content of MUFA, were associated with its preventive properties [42]. These evidences suggest that the amounts, as well as the types, of fatty acids influence cancer incidence and development [43].

It is known that the content of C18:2 n-6 in olive oils can vary significantly depending on the source and brand [44]. This variability can explain the conflicting data reported above showing different conclusions about the influence of monounsaturated fatty acids and cancer risk. In addition, these data emphasizes the need to have a broad view regarding the investigation of fatty acids intake and cancer by evaluating not only the effects of a unique fatty acid [11, 44], but the correlation of different types of fatty acids ratios. Another concern relies on the fact that olive oil contains other molecules rather than fat, such as tocopherols, carotenoids, polyphenols, that could interfere in the patterns of cancer risk [19]. Hence, there is also the need to establish which components really yield cancer prevention.

Unveiling Oleic Acid Mechanisms of Action in Cancer

The putative anticancer effects of oleic acid led to further investigations aiming to improve the comprehension of its mechanisms in a cellular and molecular basis. Some examples are described below.

Some years ago, it has been assumed that gene expression was not affected directly by nutrients. Nevertheless, fatty acids can not only be an energy source for the body, but also act as signaling molecules and mediate gene expression of genes involved in glucose and lipid metabolism [45, 46]. For example, OA is able to interfere in the gene expression of genes involved in cancer progression (*e.g.* HER-2 and FASN). The mechanisms by which it occurs have been unveiled showing promising results for cancer therapy.

HER-2 is one of the most studied genes known to be overexpressed in certain cancer types such as breast cancer. In normal cells, this gene is tightly regulated and is involved in the regulation of a diverse of biological functions as apoptosis, cell motility, differentiation, transformation, and proliferation [47]. Deregulation of HER-2 gene leading to its overexpression has been associated with oncogenic transformation, tumorigenesis, and metastasis [47]. Furthermore, patients bearing tumors displaying high expression of HER-2 usually have a poor prognosis, shorter relapse time, and low survival rate [48]. Thus, approaches aiming to target HER-2 have been intensively developed.

Surprisingly, it has been described that physiological concentrations of exogenously supplemented OA (10 μM) can markedly suppress HER-2 expression of breast cancer cells. A deeper investigation of the molecular mechanism revealed that HER-2 genes are indirectly suppressed by OA and that the protein FASN (involved in endogenous fatty acid sysnthesis) corroborates in this process. Briefly, exogenously supplemented OA down-regulate FASN promoting an accumulation of the substrate malonyl-CoA leading to a high expression of PEA3, which is an endogenous inhibitor of HER-2 transcription [12]. OA does not influence the HER-2 expression in normal cells, since the expression of HER-2 is naturally inhibited by high levels of PEA3 [49]. Interestingly, in a 7,12-dimethylbenz(a)anthracene (DMBA)-induced mammary tumorigenesis model in rats, a high olive oil diet showed to be a negative modulator conferring to the tumor a more benign clinical behavior and a lower histopathological malignancy, and this effect was accompanied by a decrease in the mRNA levels coding for HER2 [50, 51].

The rates of HER-2 suppression by OA are similar to the ones induced by the medication Trastuzumab (Herceptin®), which is the only HER-2 oncogene-targeted therapy approved by the US Food and Drug Administration (FDA) for the treatment of HER2-positive metastatic breast cancer [52]. Trastuzumab is a humanized monoclonal antibody that binds to the HER-2 ectodomain and inhibits its function. Although it has been extensively used to treat HER-2 positive tumors, the majority of patients develop resistance to the treatment within 1 yr [52]. In order to overcome this limitation, it has been proposed that the association of Trastuzumab with an inhibitor of HER 2 expression inhibitor could increase the magnitude and duration of response to Trastuzumab [52]. The rationale is that a better outcome could be obtained by preventing the transcription of 2-10 gene copies than trying to neutralize up to 10^6 HER-2 receptor molecules commonly found in HER-2-overexpressing cancer cells [53]. In fact, the combination of Trastuzumab and OA showed a strong synergism in inducing down-regulation of HER-2 and enhancing Trastuzumab efficacy towards HER-2 by promoting DNA fragmentation associated with apoptotic cell death [54].

It has been suggested that OA can induce a similar mechanism of action in various types of human malignancies, including breast, ovarian and stomach carcinomas. Moreover, other fatty acids showing protective and anti-cancer properties, such as PUFA n-3, might trigger similar mechanisms to down-regulate HER-2 expression [12]. The effects of OA in regulating HER-2 transcription can be correlated with the protective effects attributed to the "Mediterranean Diet" not only in the incidence but also in the progression (invasion and metastasis) of cancer [12].

As previously discussed, there is a need to clarify if the anticancer effects of olive oil can be attributed mainly to OA or to other molecules present in the oil. Lathan *et al* compared the effects of olive oil and OA in Caco-2 and HT-29 cells. The effects exerted by the oils was cell-line specific and by different mechanisms. OA induced apoptosis in HT-29; however, the downregulation of COX-2 and Bcl-2 expression induced by olive oil was not seen in OA-treated cells. These data suggest that other components in olive oil are involved in apoptosis induction either synergistically or in other pathways. Although OA was not able to significantly induce apoptosis in Caco-2 cells, it promoted their differentiation. The ability of substances to induce Caco-2 differentiation has been an indication of anticancer potential since this process is associated with changes in cell structure and metabolism leading to a more specialized function and to a reduction on cell proliferation [55]. Interestingly, the induction of Caco-2 differentiation was fully coincident between olive oil and OA, indicating that this effect was induced by OA [56].

Conversely, mechanisms of action showing OA tumor promoting effects have also been described. OA is able to promote proliferation, metastasis and activation of cell survival pathways (e.g. MEK1/2, ERK1/2) in breast cancer cells [57]. Investigations of mechanisms of action have shown that OA can significantly increase metastasis in MDA-MB-231 breast cancer cells. It can induce the activation of the signaling molecule FAK which is involved in the regulation of cell migration, invasion, and metastasis. A signaling pathway was delineated showing that OA promotes the production of arachidonic acid, which is metabolized by lipoxygenases; subsequently, arachidonic acid metabolites mediate FAK activation and cell migration [58]. Moreover, it was reported that OA could also promote metastasis by increasing the synthesis and secretion of metalloproteases related to extracellular matrix degradation such as MMP-9 [59].

The apparent contradictory OA effects/mechanisms should be further explored since they can vary according to the dose, time of exposure, and cell line. For example, in the MDA-MB-231 breast cancer cell line, OA was shown to be stimulatory at low (0.23-0.9 µmol/L) and inhibitory at higher (3.5 and 7.0 µmol/L) concentrations [43].

OA as Co-Adjuvant Molecules in Anticancer Approaches

Synergism with Anticancer Drugs

It has been described that fatty acids have the ability to enhance the cytotoxicity of several anticancer drugs *in vitro* and *in vivo* [60, 61]. One of the mechanisms proposed to be involved in these effects is related to biophysical and functional changes in tumor cell membranes induced by fatty acid supplementation. High ratios of unsaturated/saturated fatty acids lead to an increase of cell membrane fluidity and, consequently, enhance drug diffusion [62, 63]. Menéndez *et al*, investigated if OA was able to improve the effectiveness of paclitaxel, a known lipid soluble anticancer drug, in breast cancer cells *in vitro* by modulating cell membrane fluidity [64]. Interestingly, when both OA and paclitaxel was added to different breast cancer cell lines, only a slight improvement of paclitaxel toxicity was observed. However, when cells were pretreated with OA for 24 h prior to paclitaxel treatment, the magnitute of OA in sensitizing breast cancer cells to paclitaxel was significantly increased in a dose-dependent manner but only in MDA-MB-231 cells. Similarly, the effects of OA in increasing the sensitivity of breast cancer cells to the anticancer drug vironelbine (Navelbine®) was dependent on the treatment schedule and the type of breast cancer cell line. Thus, the data suggest that although OA effects in improving paclitaxel or vironelbine activity might be involved with changes in cell membrane fluidity, the mechanism of action seems to be more complex [64].

It seems that synergic effects of OA will vary according to the mechanism of action of the anticancer agent. For example, conjugated linoleic acid (CLA) induces cytotoxicity activity in hepatoma cells *in vitro*. During this process, CLA reduces the levels of OA by the inhibition of the enzyme delta-9 desaturase which metabolizes saturated fatty acids to monounsaturated, such as stearic acid (C18:0) to oleic acid (C18:1). In this context, co-treatment with OA showed an antagonic effect. Supplementation of OA led to an increase of OA concentrations and significantly inhibited CLA cytotoxicity [65]. Thus, further studies are necessary to better understand the synergic and antagonic effects of OA to tailor an appropriate and effective anticancer treatment.

OA as a Cofactor to Form Novel Anticancer Complexes

OA has been described to stabilize partially unfolded proteins originating novel complexes with new functions, including anticancer. Human α-lactalbumin (α-LA) made lethal to tumor cells (HAMLET) and equine lysozyme with oleic acid (ELOA) are examples of such complexes exhibiting anticancer activities markedly different from the activity of their respective proteinaceous compounds [66]. Alpha-lactalbumin (14 kDa) is a ubiquitous calcium-binding milk protein [67] that undergoes to a partially unfolded state when submitted to high pHs, high temperatures, or has its calcium chelated. This new structure is specifically stabilized by the binding of monounsaturated fatty acids (C18:1), including OA, but not other fatty acids (*e.g.* C16 or C20) suggesting a stereo-specific fit between both molecules [68]. In addition, OA was found to be a cofactor necessary not only for protein stabilization but also to form a functional anticancer complex [67, 68]. Complexes of OA with α-LA from other species (*e.g.* cow) or with equine lysozyme (ELOA), a structural homolog to α-LA, have also been shown to induce cytotoxicity to tumor cells [69, 70].

HAMLET, and also its bovine equivalent BAMLET, induces cell death within few hours of exposition in several cancer cells lines, regardless the p53 status or the Bcl-2 genotype; whereas no or low cytotoxicity is observed in normal cells [68, 71]. In tumor cells, HAMLET is translocated to the nucleus, where about 90% of HAMLET is found after 1 hour at the LD_{50} concentration. In healthy cells, HAMLET is retained in the cytoplasm [68]. Different mechanisms of action have been proposed, including autophagy, depolarization of the mitochondria, binding to histones and perturbation of chromatin structure, as well as changes in proteasome structure and function, and lysosomal membrane permeabilization [72]. Some of these effects are involved with non-apoptotic cell death pathways which have been reported to be an interesting strategy for treating highly apoptosis-resistant cancers [73].

HAMLET has been evaluated in *in vivo* studies showing promising results on bladder cancer development in mice and on human glioblastoma xenografts in the brain of nude rats with no side effects when it was administered to healthy animals [74]. Moreover, cytotoxic effects of HAMLET were observed in clinical trials on bladder cancer and skin papillomas but not in healthy surrounding tissue [75, 76]. For example, when administrated topically on papilloma lesions of patients, who had previously tested a variety of treatments without success, HAMLET reduced the lesion in almost 100% of the patients [75]. The results suggested that HAMLET should be further tested topically in other types of cancer.

The cytotoxicity of HAMLET and BAMLET has been widely evaluated; however, it is not clear whether the cytotoxicity of the complexes resides with the protein or the lipid moiety, or whether a synergistic effect of α-LA and OA is needed [72]. A recent work have reported that OA, and not the partially unfolded lactalbumin, would be the main cytotoxic component in BAMLET affecting not only cancer cells but also non-cancer according to the cell line. In the literature, there is a great variability in the ratios of OA:lactalbumin used to prepare the complexes and also in the protocols applied to measure OA concentrations [72]; thus further investigations taking into account such variables are crucial to better clarify this subject.

OA as Skin Penetration Enhancers

Topical application of anticancer drugs is another approach for cancer treatment. Depending on the tumor nature, size and local of occurrence, the transdermal delivery of drugs is an interesting option to deliver the drug specifically in the tumor region and avoid side effects that may be related to the systemic exposure of the drug [77]. However, transdermal delivery of some drugs is often impaired because the structural organization of the skin is an efficient physical barrier to the external environment. To overcome this drawback, several molecules have been studied to increase transdermal delivery of drugs. OA alone or in combination with other surfactants, polymers or fatty acids have been described to act as efficient penetration enhancers enabling the migration of drugs through the human skin. Due to its amphipathic nature, OA may be more effectively absorbed by the skin which is constituted by alternating lipophilic and hydrophilic layers [78].

The addition of OA was shown to significantly increase the permeability of Tamoxifen metabolites, used for breast cancer treatment, through the skin and also induced a favorable rapid delivery of the metabolites equivalent to that of estradiol, a widely used transdermal hormone [77]. OA has also been described to enhance skin penetration of photosensitive agents used in photodynamic therapy, which is an anticancer therapy based on the activation of photosensitive agents localized in the tumor region by a laser light leading to the production of highly reactive oxygen species that will destroy nearby tumor cells [79]. The skin accumulation of the photosensitive agent aminolevulinic acid (5-ALA), a hydrophilic molecule with limited skin penetration, was significantly optimized when OA was used [80]. Similarly, improved topical delivery of the hydrophobic photosensitive agent zinc phthalocyanine by the addition of OA was also reported [81].

OA and Nanoparticles Platforms

A great diversity of nanotechnology platforms suitable for cancer diagnostics and therapy have emerged. Nanoparticles represent an important branch of these platforms and have been a subject of extensive research, particularly for their ability to improve the efficacy of anticancer drugs [82]. Nanoparticles are able to accumulate into the tumor microenvironment by the enhanced permeability and retention (EPR) effect due to the unique physiology of cancer fenestrated vascular architecture and poor lymphatic drainage [83]. In addition, they show other interesting features for a drug delivery system such as prolonged blood circulation time, sustained drug release profile and even intracellular drug delivery [82, 83]. Owing to its amphiphilic nature, OA have been included as components of several nanoparticles as nanostructured lipid carriers (NLC), magnetic, and hydroxyapatite nanoparticles. Some applications of OA in this field are discussed in the next paragraphs.

Lipids have been largely used as the main building blocks or as a co-factor aiming to improve nanoparticle stabilization or drug load efficiency. The use of natural biomolecules in nanoparticle formulations, such as lipids, is interesting because they are expected to have low or no toxicity either *in vivo* or in the environment as a by-product [82]. Nanostructure lipid carriers (NLC) are colloidal particles composed by a mixture of solid- and liquid-phase lipids. OA is a typical liquid-phase lipid used in NCLs to control particle size and drug release. For example, as the concentration of OA is increased up to about 30%, particle size decreases, and particle morphology become more spherical, smooth, and regular [84].

NCLs have been considered attractive candidates for the delivery of anticancer agents due to their nanoscale size (100 to 500 nm), multifunctionalization sites, and high drug load

[82, 85, 86]. Moreover, NCLs can protect an anticancer drug from degradation, extend its blood circulation time and enhance treatment effects by accumulating in the tumor region via EPR effect [82, 86]. The presence of a hydrophobic core provides a suitable environment for hydrophobic drugs entrapment, which is an important advantage given that approximately 40% of newly developed drugs are hydrophobic in nature [85, 86].

The loading of the hydrophobic anticancer drug paclitaxel into NLCs have been found to overcome cell-multi drug resistance probably due to an efficient endocytosis of nanoparticles, thereby bypassing the P-glycoprotein drug efflux mechanism [82, 86].

Magnetic nanoparticles are composed by elements able to respond to a magnetic field such as iron. In cancer research, they have been intensively studied for drug targeting mediated by magnetic-field and as high performance contrast agents for magnetic resonance imaging [87, 88]. In addition, these nanoparticles play an important role in cancer magnetic hyperthermia. This therapy is characterized by subjecting magnetic nanoparticles, previously localized in the tumor region, to an alternating magnetic field resulting in production of heat that will, subsequently, destroy the tumor cells.

Magnetic nanoparticles in an aqueous solution tend to agglomerate in order to reduce their surface energy by strong dipole–dipole interactions and van der Waal forces among particles.

Therefore, further surface chemical modifications are needed to enable their use in anticancer therapies [88]. In general, surfactants or polymers have been used as crucial dispersion agents during nanoparticles synthesis. The surfactant coat can be chemically anchored or physically adsorbed on magnetic nanoparticles creating repulsive (mainly as steric repulsion) forces to balance the magnetic and the van der Waals attractive forces acting on the nanoparticles [89, 90].

The ratio of surfactant during magnetic nanoparticle synthesis is also a decisive parameter to control the size and morphology of the nanoparticles [88, 90]. OA is one of the main surfactants used during magnetic nanoparticle synthesis and it has been described as one of the best candidates for the stabilization of Fe_3O_4 in aqueous solution [90-92]. It was reported that OA is efficiently adsorbed on the Fe_3O_4 nanoparticle surface and that the total amount of adsorbed OA, as well as the density of OA layers on Fe_3O_4 nanoparticles, varies according to the size showing a decrease in the number of layers with the increase of the nanoparticle size [93].

To actively target magnetic nanoparticles to tumors, the attachment of peptides, antibodies or other targeting molecules on the surface of the nanoparticle is required. Thus, the coating layer of a nanoparticle should not only avoid nanoparticle aggregation, but should also have surface reactive moieties that allows further incorporation of active targeting moiety [88]. In this context, OA coating also yield the possibility of surface modification to enable the conjugation of functional groups (*e.g.* alkaline group), proteins, and even DNA [87, 94].

OA surface coating has also been applied to other nanoparticles rather than magnetic nanoparticles. Hydroxyapatite, the principal inorganic constituent of human bones and teeth [95, 96], constitute the core of nanoparticles showing special features such as improved biocompatibility, non-immunogenicity, non-inflammatory behavior. Although drugs can be chemically conjugated or ionically adsorbed to the surface of these nanoparticles, the drug loading capacity is low and show rapid drug release [96].

Coating the surface of hydroxyapatite nanoparticles with OA was suggested in order to improve drug loading. In this formulation, OA was used as a reservoir for hydrophobic drugs and also as a platform to anchor poliethyleneglycol moieties (DSPE-PEG-COOH) which was important to increase the nanoparticle solubility in aqueous solution. This drug delivery system was able to load significant amounts of the anticancer drug Docetaxel exhibiting a sustained release profile *in vitro*. In addition, the pharmacologic effects of Docetaxel were retained and enhancement of the therapeutic activity against hormone refractory prostate cancer was reported [96].

CONCLUSION

Nature has been an interesting source of bioactive compounds able to prevent/treat several diseases including cancer. Some of these bioactive compounds belong to classic classes of macromolecules such as lipids. In this review, the multi-facetted effects and uses of the fatty acid oleic acid on cancer were presented. OA has shown a large spectrum of actions as an anticancer agent itself (prevention and suppression) extending its role as a co-factor for other anticancer approaches such as sensitizing cancer cells and also increasing the penetration of chemotherapeutic drugs through the skin barrier. This fatty acid can also act as a stabilizing agent for proteins with anticancer activity. As a component of nanoparticles, OA helps to control particle aggregation, size, and is also suitable for chemical modifications enabling the attachment of target moieties on the nanoparticle surface. Despite the interesting effects described above, there are still controversial data regarding OA beneficial effects in cancer. Therefore, further investigations of OA nutritional epidemiology and mechanisms of action are desired in order to provide a solid basis for consolidating its future clinical settings in anticancer approaches.

REFERENCES

[1] World Health Organization, 2008 at http://www.who.int/topics/ global_burden_of_disease/en/
[2] World Health Organization, 2011 at http://www.who.int/cancer/en/
[3] World Health Organization Databank, 2010 at http://www.who.int/ whosis.
[4] D. Hanahan and R. A. Weinberg, *Cell* 144, 646 (2011).
[5] F. Azam, S. Mehta and A. L. Harris, *Eur. J. Cancer* 46, 1323 (2011).
[6] G. Danaei, S. Vander Hoorn, A. D. Lopez, C. J. Murray and M. Ezzati, *Lancet* 366, 1784 (2005).
[7] L. N. Kolonel, D. Altshuler and B. E. Henderson, *Nat. Rev. Cancer* 4, 519 (2004).
[8] R. Doll, *Cancer Res.* 52, 2024s (1992).
[9] P. Buell, *J. Natl. Cancer Inst.* 51, 1479 (1973).
[10] R. G. Ziegler, R. N. Hoover, M. C. Pike, A. Hildesheim, A. M. Nomura, D. W. West, A. H. Wu-Williams, L. N. Kolonel, P. L. Horn-Ross, J. F. Rosenthal and M. B. Hyer, *J. Natl. Cancer Inst.* 85, 1819 (1993).
[11] P. Bougnoux, B. Giraudeau and C. Couet, *Cancer Epidemiol. Biomarkers Prev.* 15, 416 (2006).
[12] J. A. Menendez and R. Lupu, *Curr. Pharm. Biotechnol.* 7, 495 (2006).

[13] M. Gerber, *Cancer Lett.* 114, 91 (1997).

[14] Trichopoulou, T. Costacou, C. Bamia and D. Trichopoulos, *N. Engl. J. Med.* 348, 2599 (2003).

[15] Keys, C. Aravanis and H. Van-Buchem, *Lancet* 2, 58 (1981).

[16] R. W. Owen, A. Giacosa, W. E. Hull, R. Haubner, G. Wurtele, B. Spiegelhalder and H. Bartsch, *Lancet Oncol.* 1, 107 (2000).

[17] S. A. Bingham, R. Luben, A. Welch, N. Wareham, K. T. Khaw and N. Day, *Lancet* 362, 212 (2003).

[18] V. Chajes and P. Bougnoux, *World Rev. Nutr. Diet* 92, 133 (2003).

[19] J. M. Martin-Moreno, W. C. Willett, L. Gorgojo, J. R. Banegas, F. Rodriguez-Artalejo, J. C. Fernandez-Rodriguez, P. Maisonneuve and P. Boyle, *Int. J. Cancer* 58, 774 (1994).

[20] Trichopoulou, K. Katsouyanni, S. Stuver, L. Tzala, C. Gnardellis, E. Rimm and D. Trichopoulos, *J. Natl. Cancer Inst.* 87, 110 (1995).

[21] Trichopoulou, P. Lagiou, H. Kuper and D. Trichopoulos, *Cancer Epidemiol. Biomarkers Prev.* 9, 869 (2000).

[22] L. E. Voorrips, H. A. Brants, A. F. Kardinaal, G. J. Hiddink, P. A. van den Brandt and R. A. Goldbohm, *Am. J. Clin. Nutr.* 76, 873 (2002).

[23] M. Saadatian-Elahi, T. Norat, J. Goudable and E. Riboli, *Int. J. Cancer* 111, 584 (2004).

[24] R. L. Prentice, *J. Natl. Cancer Inst.* 88, 1738 (1996).

[25] E. Riboli, H. Ronnholm and R. Saracci, *Cancer Surv.* 6, 685 (1987).

[26] R. M. Dougherty, C. Galli, A. Ferro-Luzzi and J. M. Iacono, *Am. J. Clin. Nutr.* 45, 443 (1987).

[27] B. Wood, N. A. Habib, A. Thompson, H. Bradpiece, C. Smadja, M. Hershman, W. Barker and K. Apostolov, *Br. Med. J. (Clin. Res. Ed.)* 291, 163 (1985).

[28] M. Pandey, A. K. Khatri, S. S. Dubey, A. Gautam and V. K. Shukla, *Eur. J. Surg. Oncol.* 24, 43 (1998).

[29] S. Thomas, J. O'Dea and I. S. Fentiman, *Br. J. Surg.* 75, 1078 (1988).

[30] J. P. Neoptolemos and B. S. Thomas, *Ann. Clin. Biochem.* 27, 38 (1990).

[31] S. Dayton, S. Hashimoto, W. Dixon and M. L. Pearce, *J. Lipid Res.* 7, 103 (1966).

[32] L. Kohlmeier and M. Kohlmeier, *Environ. Health Perspect.* 103, 99 (1995).

[33] R. Bartoli, F. Fernandez-Banares, E. Navarro, E. Castella, J. Mane, M. Alvarez, C. Pastor, E. Cabre and M. A. Gassull, *Gut* 46, 191 (2000).

[34] S. Reddy and Y. Maeura, *J. Natl. Cancer Inst.* 72, 745 (1984).

[35] L. A. Cohen, D. O. Thompson, Y. Maeura, K. Choi, M. E. Blank and D. P. Rose, *J. Natl. Cancer Inst.* 77, 33 (1986).

[36] M. P. Fay, L. S. Freedman, C. K. Clifford and D. N. Midthune, *Cancer Res.* 57, 3979 (1997).

[37] W. Nakahara, *J. Exp. Med.* 41, 347 (1925).

[38] K. K. Carroll and H. T. Khor, *Lipids* 6, 415 (1971).

[39] J. Tinsley, J. A. Schmitz and D. A. Pierce, *Cancer Res.* 41, 1460 (1981).

[40] B. Katz and E. S. Boylan, *Nutr. Cancer* 12, 343 (1989).

[41] M. S. Wicha, L. A. Liotta and W. R. Kidwell, *Cancer Res.* 39, 426 (1979).

[42] B. Lasekan, M. K. Clayton, A. Gendron-Fitzpatrick and D. M. Ney, *Nutr. Cancer* 13, 153 (1990).

[43] P. Rose, *Am. J. Clin. Nutr.* 66, 998S (1997).
[44] Ip, *Am. J. Clin. Nutr.* 66, 1523S (1997).
[45] M. J. McArthur, B. P. Atshaves, A. Frolov, W. D. Foxworth, A. B. Kier and F. Schroeder, *J. Lipid Res.* 40, 1371 (1999).
[46] B. Jump and S. D. Clarke, *Ann. Rev. Nutr.* 19, 63 (1999).
[47] Y. Yarden and M. X. Sliwkowski, *Nat. Rev. Mol. Cell Biol.* 2, 127 (2001).
[48] Zhang, L. K. Tai, L. L. Wong, L. L. Chiu, S. K. Sethi and E. S. Koay, *Mol. Cell Proteomics* 4, 1686 (2005).
[49] A. Menendez, A. Papadimitropoulou, L. Vellon and R. Lupu, *Eur. J. Cancer* 42, 2425 (2006).
[50] R. Moral, M. Solanas, G. Garcia, R. Colomer and E. Escrich, *Oncol. Rep.* 10, 1417 (2003).
[51] Solanas, A. Hurtado, I. Costa, R. Moral, J. A. Menendez, R. Colomer and E. Escrich, *Int. J. Oncol.* 21, 745 (2002).
[52] R. Nahta, D. Yu, M. C. Hung, G. N. Hortobagyi and F. J. Esteva, *Nat. Clin. Pract. Oncol.* 3, 269 (2006).
[53] S. C. Wang and M. C. Hung, *Drugs Today* 36, 835 (2000).
[54] J. A. Menendez, S. Ropero, R. Lupu and R. Colomer, *Ann. Oncol.* 15, 1719 (2004).
[55] P. Latham, E. K. Lund and I. T. Johnson, *Carcinogenesis* 20, 645 (1999).
[56] X. Llor, E. Pons, A. Roca, M. Alvarez, J. Mane, F. Fernandez-Banares and M. A. Gassull, *Clin. Nutr.* 22, 71 (2003).
[57] Soto-Guzman, T. Robledo, M. Lopez-Perez and E. P. Salazar, *Mol. Cell Endocrinol.* 294, 81 (2008).
[58] Navarro-Tito, A. Soto-Guzman, L. Castro-Sanchez, R. Martinez-Orozco and E. P. Salazar, *Int. J. Biochem. Cell. Biol.* 42, 306 (2010).
[59] Soto-Guzman, N. Navarro-Tito, L. Castro-Sanchez, R. Martinez-Orozco and E. P. Salazar, *Clin. Exp. Metastasis* 27, 505 (2010).
[60] Germain, V. Chajes, S. Cognault, C. Lhuillery and P. Bougnoux, *Int. J. Cancer* 75, 578 (1998).
[61] Y. Shao, L. Pardini and R. S. Pardini, *Lipids* 30, 1035 (1995).
[62] Timmer-Bosscha, G. A. Hospers, C. Meijer, N. H. Mulder, F. A. Muskiet, I. A. Martini, D. R. Uges and E. G. de Vries, *J. Natl. Cancer Inst.* 81, 1069 (1989).
[63] J. G. Zijlstra, E. G. de Vries, F. A. Muskiet, I. A. Martini, H. Timmer-Bosscha and N. H. Mulder, *Int. J. Cancer* 40, 850 (1987).
[64] J. A. Menendez, M. del Mar Barbacid, S. Montero, E. Sevilla, E. Escrich, M. Solanas, H. Cortes-Funes and R. Colomer, *Eur. J. Cancer* 37, 402 (2001).
[65] Yamasaki, H. Chujo, S. Nou, H. Tachibana and K. Yamada, *Cancer Lett.* 196, 187 (2003).
[66] K. Mossberg, K. Hun Mok, L. A. Morozova-Roche and C. Svanborg, *FEBS J.* 277, 4614 (2010).
[67] M. Svensson, A. Hakansson, A. K. Mossberg, S. Linse and C. Svanborg, *Proc. Natl. Acad. Sci. U S A* 97, 4221 (2000).
[68] Hallgren, S. Aits, P. Brest, L. Gustafsson, A. K. Mossberg, B. Wullt and C. Svanborg, *Adv. Exp. Med. Biol.* 606, 217 (2008).
[69] V. Vukojevic, A. M. Bowen, K. Wilhelm, Y. Ming, Z. Ce, J. Schleucher, P. J. Hore, L. Terenius and L. A. Morozova-Roche, *Langmuir* 26, 14782 (2010).

[70] J. Pettersson, A. K. Mossberg and C. Svanborg, *Biochem. Biophys. Res. Commun.* 345, 260 (2006).

[71] R. Brinkmann, S. Thiel, M. K. Larsen, T. E. Petersen, J. C. Jensenius and C. W. Heegaard, *J. Dairy Sci.* 94, 2159 (2011).

[72] R. Brinkmann, C. W. Heegaard, T. E. Petersen, J. C. Jensenius and S. Thiel, *FEBS J.* 278, 1955 (2011).

[73] Rammer, L. Groth-Pedersen, T. Kirkegaard, M. Daugaard, A. Rytter, P. Szyniarowski, M. Hoyer-Hansen, L. K. Povlsen, J. Nylandsted, J. E. Larsen and M. Jaattela, *Mol. Cancer Ther.* 9, 24 (2011).

[74] W. Fischer, L. Gustafsson, A. K. Mossberg, J. Gronli, S. Mork, R. Bjerkvig and C. Svanborg, *Cancer Res.* 64, 2105 (2004).

[75] L. Gustafsson, I. Leijonhufvud, A. Aronsson, A. K. Mossberg and C. Svanborg, *N. Engl. J. Med.* 350, 2663 (2004).

[76] K. Mossberg, B. Wullt, L. Gustafsson, W. Mansson, E. Ljunggren and C. Svanborg, *Int. J. Cancer* 121, 1352 (2007).

[77] O. Lee, D. Ivancic, R. T. C. Jr, A. W. Rademaker and S. A. Khan, *Breast Cancer: Targets and Therapy* 3 61 (2011).

[78] C. Guillard, C. Laugel and A. Baillet-Guffroy, *Eur. J. Pharm. Sci.* 36, 192 (2009).

[79] E. Dolmans, D. Fukumura and R. K. Jain, *Nat. Rev. Cancer* 3, 380 (2003).

[80] M. B. Pierre, E. Ricci, Jr., A. C. Tedesco and M. V. Bentley, *Pharm. Res.* 23, 360 (2006).

[81] R. da Silva, Z. M. de Freitas, B. Gitirana Lde and E. Ricci-Junior, *Drug Dev. Ind. Pharm.* 37, 569 (2011).

[82] Puri, K. Loomis, B. Smith, J. H. Lee, A. Yavlovich, E. Heldman and R. Blumenthal, *Crit. Rev. Ther. Drug Carrier Syst.* 26, 523 (2009).

[83] V. Torchilin, *Adv. Drug. Deliv. Rev.* 63, 131 (2010).

[84] Q. Hu, S. P. Jiang, Y. Z. Du, H. Yuan, Y. Q. Ye and S. Zeng, *Int. J. Pharm.* 314, 83 (2006).

[85] S. Martins, B. Sarmento, D. C. Ferreira and E. B. Souto, *Int. J. Nanomedicine* 2, 595 (2007).

[86] X. G. Zhang, J. Miao, Y. Q. Dai, Y. Z. Du, H. Yuan and F. Q. Hu, *Int. J. Pharm.* 361, 239 (2008).

[87] M. A. White, J. A. Johnson, J. T. Koberstein and N. J. Turro, *J. Am. Chem. Soc.* 128, 11356 (2006).

[88] Qiao, C. Yang and M. Gao, *J. Mater. Chem.* 19, 6274 (2009).

[89] L. Shen, P. E. Laibinis and T. A. Hatton, *Langmuir* 15, 447 (1999).

[90] A.-H. Lu, E. L. Salabas and F. Schuth, *Angew. Chem. Int.* 46, 1222 (2007).

[91] L. Willis, N. J. Turro and S. O'Brien, *Chem. Mater.* 17, 5970 (2005).

[92] L. Cushing, V. L. Kolesnichenko and C. J. O'Connor, *Chem. Rev.* 104, 3893 (2004).

[93] L. Zhang, R. He and H.-C. Gu, *Applied Surface Science* 253, 2611 (2006).

[94] G. Grancharov, H. Zeng, S. Sun, S. X. Wang, S. O'Brien, C. B. Murray, J. R. Kirtley and G. A. Held, *J. Phys. Chem. B* 109, 13030 (2005).

[95] Shen, X. Hu, F. Yang, J. Bei and S. Wang, *Acta Biomater.* 6, 455 (2010).

[96] Y. Luo, Y. Ling, W. Guo, J. Pang, W. Liu, Y. Fang, X. Wen, K. Wei and X. Gao, *J. Control Release* 147, 278 (2010).

INDEX

A

B

C

D

E

F

G

H

I

J

K

L

M

N

O

Q

R

T